THE ULTIMATE GUIDE TO ABOVE-GROUND POOLS

THE ULTIMATE GUIDE TO ABOVE-GROUND POOLS

Terry Tamminen

McGraw-Hill

New York   Chicago   San Francisco   Lisbon   London   Madrid
Mexico City   Milan   New Delhi   San Juan   Seoul
Singapore   Sydney   Toronto

The McGraw·Hill Companies

Cataloging-in-Publication Data is on file with the Library of Congress

Copyright © 2004 by The McGraw-Hill Companies, Inc. All rights reserved. Printed in the United States of America. Except as permitted under the United States Copyright Act of 1976, no part of this publication may be reproduced or distributed in any form or by any means, or stored in a data base or retrieval system, without the prior written permission of the publisher.

1 2 3 4 5 6 7 8 9 0   DOC/DOC   0 1 0 9 8 7 6 5 4

ISBN 0-07-142514-4

The sponsoring editor for this book was Larry S. Hager, the editing supervisor was Stephen M. Smith, and the production supervisor was Pamela A. Pelton. It was set in Melior by Vicki Hunt of McGraw-Hill Professional's Hightstown, N.J., composition unit. The art director for the cover was Anthony Landi.

Printed and bound by RR Donnelley.

McGraw-Hill books are available at special quantity discounts to use as premiums and sales promotions, or for use in corporate training programs. For more information, please write to the Director of Special Sales, McGraw-Hill Professional, Two Penn Plaza, New York, NY 10121-2298. Or contact your local bookstore.

 This book is printed on recycled, acid-free paper containing a minimum of 50% recycled, de-inked fiber.

To the men and women of the Waterkeeper Alliance, for making our water safe for drinking, swimming, and passing on to the next generation.

CONTENTS

ACKNOWLEDGMENTS

It is impossible to assemble as much detailed technical information as is presented in this volume without considerable assistance from many people, especially the technical departments at pool and equipment manufacturers. My sincere appreciation and thanks to everyone who contributed.

By using one manufacturer's product as an example for each topic, I have tried to show the products you will encounter most often in the field. My excluding other products is not a suggestion that they are not the equals of the ones shown, just the reality that I had to choose a representative item in each case. Keep an open mind about new products or manufacturers adding new items to their lines; you'll find great new improvements and values.

I owe thanks to just about every company in the water technology industry for their training, product innovations, product literature, and advice. The following list reflects my special gratitude to those who made specific contributions to the book:

- Kris Haddad, organizer extraordinaire!

- Aqua-Flo, Inc., Chino, California; Lauren Caudill.

- Artesian Pools, Baltimore, Maryland; Greg Fink.

- Cantar, a division of Cantar/Polyair Corp., Youngstown, Ohio; Brian Frost.

- Cornelius Pools, LLC, Rancho Cucamonga, California; Brad Rinehart.

- Delair Group, LLC, Delair, New Jersey; Ilene Fink.

- FeherGuard Products Ltd., Milton, Ontario, Canada; Mike Harvey.

- Franklin Electric, Bluffton, Indiana; Mel Haag.

- Gladon Co., Milwaukee, Wisconsin; Thomas J. Haight.

- Hanna Instruments, Woonsocket, Rhode Island; Detmar Finke.

- Hoffinger Industries/Doughboy Pools, Southaven, Mississippi; James L. Palmer, Randy Howard, and Dorreen Davis.

- Intermatic, Inc., Spring Grove, Illinois; Wayne Veatch.

- Jacuzzi Brothers, Little Rock, Arkansas; Lisa Poe.

- Jandy, Petaluma, California; Scott Ferguson.

- Lass Enterprises, Altamonte Springs, Florida; Alan D. Vander Boegh.

- Loop-Loc Ltd., Hauppauge, New York; Jeanne Mitchell.

- Mr. Pool, Van Nuys, California; Josh and Brian.

- North West Wholesale (discount swimming pools and supplies), Winnipeg, Manitoba, Canada; www.northwestwholesale.com, (800) 684-5494; Dan Goudreau.

- Pentair Pool Products, Inc., Moorpark, California; Robin Viola.

- Polaris Pool Systems, Inc., Vista, California; Janice Tegman.

- Pool Tool Co., Ventura, California; Laurie Gaynor and Herb Tilsner.

- Premier Spring Water, Pacoima, California; Paul R. Blum.

- Raypak, Inc., Westlake Village, California; Terry S. Doyle.

- Sharkline Pools, Hauppauge, New York; Richard Sobel.

- SmartPool, Inc., Lakewood, New Jersey; Stephen Shulman.

- Sofpool, LLC, Rancho Cordova, California; Dwayne Carreau.

- Splash SuperPools, LLC, North Little Rock, Arkansas; Tommy Thompson.

- Sta-Rite Industries, Delevan, Wisconsin; Tom Elsner.

- Summit-USA, Inc., a swimming pool slides manufacturer located in Washington State; (360) 636-4433; Leo Lin.

- Technobois, La Prairie, Quebec, Canada; Simon Tardif.

- Zodiac American Pools, Inc., Midway, Georgia; Richard Raffaelli.

As the "Malibu Poolman to the Stars," my celebrity clients have included Barbra Streisand, Dustin Hoffman, Bruce Willis, and Madonna. Ten years ago, not one of them owned an above-ground pool. Today, everyone from movie stars to mail carriers have discovered the amazing features and benefits of above-ground pools.

My pool service and construction background has encompassed everything from inflatables to $1 million artificial-rockscape water wonderlands, but if I were buying a pool today, it would be an above-ground model. Why? First, you can buy an above-ground pool that meets just about every swimming need, from spas and lap pools to family fun pools and even large commercial varieties, for a fraction of the cost of an inground model. Second, above-ground pools can move when you do, and if your swimming needs change, you can upsize or downsize accordingly, something that isn't practical once you have poured concrete and steel into a hole in your yard. Finally, as a service technician I can attest to the fact that above-ground pools are significantly easier and less costly to maintain, especially when it comes to finding and repairing leaks.

This book is the first of its kind for above-ground pools and stands on the shoulders of my earlier books, including *The Ultimate Pool Maintenance Manual* (McGraw-Hill). That book is still the only comprehensive published text on the subject of swimming pool and spa repair, maintenance, and water chemistry. Now, *The Ultimate Guide to Above-Ground Pools* offers the same level of detail, for the pool professional and do-it-yourself pool owner alike, on the fastest-growing segment of the pool industry. From the moment you open these pages, you will understand your pool better and save money on service, repairs, remodeling, and routine maintenance. You might also discover you enjoy daunting subjects like water chemistry when you understand more about what's going on below the waterline.

This book covers every type of above-ground pool and includes trade secrets, labor-saving tips, and the latest technology. Among the features you will find are:

- An Easy, Advanced, or Pro rating for each installation, repair, and service job.

- Quick Start Guides for common tasks, to help you evaluate the complexity of the work in advance and to serve as a simple reminder checklist for those who are already familiar with the skills, tools, and components involved.

- Tricks of the Trade that reveal the time- and labor-saving secrets of the pros, in print here for the first time anywhere.

- A detailed glossary that explains both common terms and confusing ones.

- A comprehensive list of the most useful websites for above-ground pools, many with informative links to manufacturers, money-saving online merchants, and exciting new products.

- The best chlorine alternatives and ways to keep your pool clean and healthful all year.

- Solar heating systems that anyone can afford and install in minutes.

- All measurements and calculations presented in both the U.S. Customary system and metric (except when describing some manufacturers' products, which may be produced in only one system).

By the way, don't let the size of this book intimidate you, because you can turn to any chapter and tackle a service or repair job in minutes by following the steps and guidelines provided. As a service technician, I've written much of the book from that perspective, but you will obtain the same results as the pros by mastering the basic skills described in the book and then applying them to each step of the pool tasks you perform.

One last word, about other sources of advice: Websites, product brochures, trade association guidelines, service technicians, and the counter staff at the pool supply store will all have advice about the "correct" way to accomplish pool maintenance. Some of that advice may conflict, not because it's wrong but because there are so many styles of above-ground pools, used under a wide variety of conditions. As a remedy, this book presents the best, field-tested techniques that will apply to the vast majority of situations. Nevertheless, every pool is unique. Take note of what works best for your pool and remember that the best teachers are common sense and asking questions (most of which can be answered in the pages of this book!).

So go right in—the water's fine! Good luck and enjoy your above-ground pool.

Terry Tamminen

The Above-Ground Pool

As its title implies, this is certainly a book about the growing variety of above-ground pools, but it is also a book about water. If you wanted to be a banker, it would be nice to understand something about the bank, vault, and cash drawers, but the real business is the money and how it is used. Similarly, this book will contain appropriate information about the "containers" because many of us will assemble our own above-ground pool, and some will take it down in winter and rebuild it every spring, so there is a lot to know. But this book will also teach the essentials of water and the related products and equipment that move or change it.

How It Works

Let's begin with a basic overview of the typical container and water system. Figure 1-1 shows a typical above-ground pool and its related equipment. Just as in banking where you follow the money if you want to learn that system, in a pool you must follow the water. That is also how this chapter (and the entire book) is outlined—in the logical pattern that the water travels from pool through plumbing to the pump/motor, filter, heater, and back to the pool.

Follow the arrows in Fig. 1-1 to follow the path of the water. The water enters the plumbing through a surface skimmer and, in some

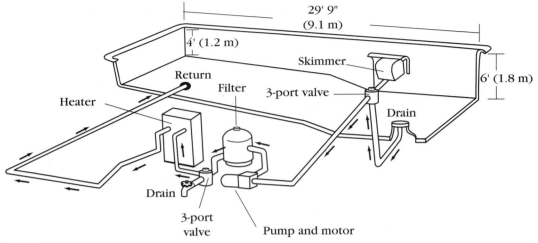

FIGURE 1-1 Typical above-ground pool with equipment.

pools, a main drain. It does this thanks to suction created by a pump and motor. After passing through the pump, the water is cleansed by a filter, warmed by a heater, and returned to the pool through a return outlet.

How Much Water Does It Hold?

Because many of the calculations in this book depend on knowing the quantity of water involved, here is how to calculate the volume of your pool.

Square or Rectangular

The formula is simple:

Length × width × average depth × 7.5 = volume (in gallons)

Let's first examine the parts of the formula. Length times width gives the surface area of the pool. Multiplying that by the average depth gives the volume in cubic feet. Since there are 7.5 gallons in each cubic foot, you multiply the cubic feet of the pool by 7.5 to arrive at the volume of the pool (expressed in gallons).

The same formula works in metric:

Length × width × average depth = cubic meters of volume

That's as far as you will probably need to take the equation since things like pool chemical dosages in metric measurements will be based on a certain amount per cubic meter. But in a smaller pool, dosages may be expressed in certain amounts per liter of water. Since there are 1000 liters in 1 cubic meter, the formula becomes

Length × width × average depth × 1000 = volume (in liters)

The formula is simple and so is the procedure. Measure the length, width, and average depth of the pool, rounding each measurement off to the nearest foot or percentage of 1 foot. If math was not your strong suit in school, remember 1 inch equals 0.0833 foot. Therefore, multiply the number of inches in your measurements by 0.0833 to get the appropriate percentage of 1 foot. Example:

$$29 \text{ ft, 9 in.} = 29 \text{ ft} + (9 \text{ in.} \times 0.0833)$$
$$= 29 + 0.75$$
$$= 29.75 \text{ ft}$$

Since metric units are already based on a decimal system in units of 10, there is no need to convert anything. As with standard units, round off to the nearest decimal for ease of calculation.

The depth of most above-ground pools is the same throughout the pool. For those with a deep end, the average depth will be only an estimate, but if the shallow end is 4 feet and the deep end is 6 feet, and assuming the slope of the pool bottom is gradual and even, then the average depth is about 5 feet. If most of the pool is only 3 or 4 feet and then a small area drops off suddenly to 6 feet, you will have a different average depth. In such a case, you might want to treat the pool as two parts. Measure the length, width, and average depth of the shallow section, and then take the same measurements for the deeper section. Calculate the volume of the shallow section and add that to the volume you calculate for the deeper section.

In either case, be sure to use the actual water depth in your calculations, not the depth of the container. For example, the shallow end of the pool as depicted in Fig. 1-2 is 4 feet deep, but the water is filled to

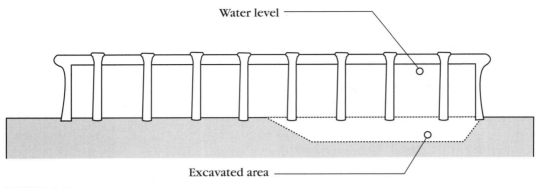

Water level

Excavated area

FIGURE 1-2 Cross-section of a typical above-ground pool.

only about 3 feet. Using 4 feet in this calculation will result in a volume 33 percent greater than the actual amount of water. This could mean serious errors when adding chemicals, for example, which are administered based on the volume of water in question. There might be a time when you want to know the potential volume, if the pool is filled to the brim. Then, of course, you would use the actual depth (or average depth) measurement. In the example, that was 4 feet.

Try to calculate the volume of the pool in Fig. 1-1:

Length × width × average depth × 7.5 = volume (in gallons)

29.75 ft × 10 ft × 5 ft × 7.5 = 11,156 gal

9.1 m × 3 m × 1.5 m × 1000 = 40,950 L (or 41 kL)

Circular

The formula:

3.14 × radius squared × average depth × 7.5 = volume (in gallons)

The calculations in metric units will be the same, except remember to multiply by 1000 instead of 7.5.

The first part, 3.14, refers to pi, which is a mathematical constant. It doesn't matter why it is 3.14 (actually the exact value of pi cannot be calculated but who cares?). For our purposes, we need only accept this as fact.

The radius is one-half the diameter, so measure the distance across the broadest part of the circle and divide it in half to arrive at the radius. Squared means multiplied by itself, so multiply the radius by itself. For example, if you measure the radius as 5 feet, multiply 5 feet by 5 feet to arrive at 25 feet. The rest of the equation was explained in the square or rectangular calculation.

Try to calculate the volume of a round container. Let's do the tricky part first. Assume the diameter of the pool is 10 feet. Half of that is 5 feet. Squared (multiplied by itself) means 5 feet times 5 feet equals 25 square feet. Assume a depth of 3 feet. Knowing this, you can return to the formula:

$$3.14 \times \text{radius squared} \times \text{average depth} \times 7.5 = \text{volume (in gallons)}$$

$$3.14 \times 25 \text{ ft} \times 3 \text{ ft} \times 7.5 = 1766.25 \text{ gal}$$

In metric, the radius of the same pool measures 1.52 meters. Multiplied by itself, this equals 2.3 meters. The average depth is 0.9 meter, so the equation looks like this:

$$3.14 \times 2.3 \text{ m} \times 0.9 \text{ m} \times 1000 = 6500 \text{ L}$$

Irregular Shapes

To calculate the capacity of irregular shapes, imagine the pool as a combination of smaller, regular shapes (Fig. 1-3). Measure these various areas and use the calculations described previously for each square or rectangular area and for each circular area. Add these volumes together to determine the total capacity. Figure 1-3 contains one rectangle and one circle (shown in two halves).

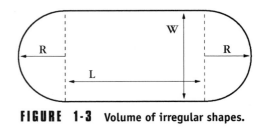

FIGURE 1-3 Volume of irregular shapes.

Parts per Million

One other important calculation you will use is parts per million (ppm). The amount of solids and liquids in the water is measured in parts per million, as in 3 parts of chlorine in every 1 million parts of water (or 3 ppm). However, 1 gallon of chlorine, for example, poured into 1 million gallons of water does not equal 1 ppm. That is because the two liquids are not of equal density. This becomes obvious when you discover that a gallon of water weighs 8.3 pounds (3.8 kilograms) but a gallon of chlorine weighs 10 pounds (4.5 kilograms) in a 15 percent solution, as described later. The chlorine is a more dense liquid—there's more of it than an equal volume of water.

To calculate parts per million, use the following example:

1 gal of chlorine in 25,000 gal of water

= 10 lb of chlorine in 207,500 lb of water

Now dividing each by 10 gives you:

1 lb of chlorine in 20,750 lb of water

So you see that 1 part of chlorine is in each 20,750 parts of water. But how does that translate to parts of chlorine per 1 million parts of water? To learn that, you must find out how many 20,750s there are in a million.

$$1,000,000 \div 20,750 = 48.19$$

$$48.19 \times 1 \text{ part of chlorine} = 48.19$$

There are 48.19 parts of chlorine in each million parts of water, expressed as 48.19 ppm.

Using the same formula without first translating the two liquids into pounds would give an answer of 40 ppm. Obviously this great discrepancy can result in substantial errors in treating water chemistry problems. But we're not through just yet.

If chlorine were 100 percent strength as it comes out of the bottle, that would be all there is to this calculation. As you will see in later chapters, that is not the case. In fact, liquid chlorine is produced in 10 to 15 percent solution, meaning 10 to 15 percent of what comes out of the bottle is chlorine and the rest is filler. Therefore, to really know how many parts of chlorine are in each million parts of water, you must adjust your result for the real amount of chlorine. Usually liquid chlorine is 15 percent strength (common laundry bleach is the same product, but around 3 percent strength), so:

$$48.19 \times 0.15 = 7.23 \text{ ppm}$$

Therefore, 7.23 ppm is our true chlorine strength in the example of 25,000 gallons (94.6 cubic meters or 94,625 liters) of water.

How Is It Made?

Ever since the invention of the creekside swimming hole, complete with swinging rope or tire, pool builders have invented new and creative ways to capture water in our backyards. Today, because of modern materials, engineering, and building techniques, there are countless types of above-ground pools.

There are essentially two categories of above-ground pools—above-ground and onground. It should be noted that manufacturers and parts providers are inconsistent with their use of these terms as well as descriptors like *portable*, *raised*, *frameless*, and *inflatable*. A standardized definition that might be helpful is to consider *above-ground* to mean structures of various sturdy frames and panels, lined with a vinyl bag, and *onground* to mean one-piece pools of various fabrics (significantly thicker than the vinyl liner of an above-ground pool) that are self-standing units, including those that are inflated. Both above-ground and onground types are portable inasmuch as they can be readily disassembled, moved, and reassembled. Since the distinction between above-ground and onground pools is most relevant in the initial purchasing decision, the chapters after this one will refer to both types as "above-ground."

Above-ground pools consist of metal (steel or aluminum), plastic, or wood frames that are assembled on top of the ground (Fig. 1-4). Prefabricated panels of similar materials are joined to the frame to create a shell, which is then lined with a heavy-duty vinyl liner to create the actual pool. Many units are made as do-it-yourself backyard kits consisting of self-supporting aluminum or steel walls and braces with a vinyl liner, ladder or stairs, decks, and equipment all packaged together. More elaborate models include a deeper end, like an inground pool, which requires some excavation of the site (Fig. 1-5).

The onground pool, which is essentially a single-piece fabric shell that simply sits on the ground, comes in two varieties. The first is the familiar inflatable pool. Zodiac, makers of the famous inflatable boat, and other manufacturers are now applying PVC-vinyl technology to pools and spas (Fig. 1-6). A variation on this type of pool is a vinyl bag that stands on its own when filled with water (Fig. 1-7A). One manufacturer of this type of pool, Sofpool, advertises the strength of its products by driving a truck into a fully filled pool (Fig. 1-7B). Amazingly,

FIGURE 1-4 Typical above-ground pools. *Delair Group, LLC.*

FIGURE 1-5 **Above-ground pool with deep end.** *Hoffinger Industries/Doughboy Pools.*

the pool bounces right back and remains undamaged. This pool is also sold in an easy-to-assemble kit with all of the equipment and accessories included (Fig. 1-8). The second variety of onground pool is designed essentially as vinyl boxes of water, supported by a framework (Fig. 1-9).

Above-ground and onground pools of all types are made in a wide variety of shapes, sizes, and depths. Common shapes are round or oval and sizes are typically 48 or 52 inches (122 or 132 centimeters) deep and up to 40 feet (12.2 meters) in diameter or length. Some pools offer a deep end option, providing a maximum depth of as much as 8 feet (2.4 meters).

You will find variations among different manufacturers, but understanding these two basic types will make it easy to understand the construction methods of all above-ground pools and help you choose the one that meets your needs. The following descriptions provide additional detail about the materials and engineering of today's above-ground and onground pools.

Above-Ground

All metal above-ground pools employ a variety of metals, plastics, and resins within the same pool unit. Manufacturers use aluminum walls for light weight and rust resistance; steel for bracing; plastic for edges,

FIGURE 1-6 Inflatable onground pools. *Zodiac American Pools, Inc.*

seats, and coatings; and stainless steel hardware for strength and durability. That said, each manufacturer's pools can be generally categorized by the predominant material used in their products.

ALUMINUM

Aluminum has the benefit of being strong and rust resistant, and therefore very long lasting. It is also lightweight, making it easier to install and, if necessary, disassemble and move. One manufacturer points out that it was these characteristics that led to the use of aluminum scaffolding during the recent renovations to the Statue of Liberty and Washington Monument.

FIGURE 1-7 (A) Frameless onground pool. (B) Strength of frameless onground pool. *Sofpool, LLC.*

FIGURE 1-8 Onground pool kit. *Sofpool, LLC.*

The framework for an aluminum pool is made by the extrusion process. Hot, soft metal is squeezed through a precut die, in the same way that toothpaste is squeezed from a tube. This method allows the resulting beams and uprights to be thicker where most strength is needed. In another method of crafting these framework components, called *roll-forming*, they are made from coiled sheets of aluminum, like the cardboard roller inside the toilet paper. Some framework requires an A shape (Fig. 1-10) to stiffen the walls sufficiently to hold the weight of the water within the vessel. Other designs, called *narrow* or *compact* buttress systems by some designers, are vertical I beams (Fig. 1-11) that perform the same function but take up less space around the pool.

The wall panels for aluminum pools are also made of either extruded or sheet metal, and they are often embossed (Fig. 1-12) and covered with decorative coatings in a wide

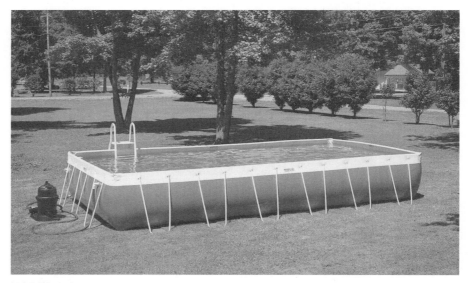

FIGURE 1-9 Typical soft-sided/framed pool. *Splash SuperPools, LLC.*

FIGURE 1-10 A-frame construction.

FIGURE 1-11 I-beam construction.

variety of colors. The walls and framework are assembled with stainless steel fasteners and brackets, chosen for even greater strength and durability at high-stress connection points. Wall panels are also prefabricated to interlock, making it easier to craft pools of various dimensions and shapes.

The vessel that actually holds the water is the vinyl liner, draped inside the metal frame and wall unit. The vinyl liner is normally 18 to 25 millimeters thick (even in the United States, where the English standard measurement is king, liners are described by their thickness in millimeters!), and it comes in every color and pattern imaginable and is often embossed or otherwise textured for cosmetic purposes. The pool in Fig. 1-13 appears to be covered entirely in small tiles, but it is actually a printed liner. The liner is attached to the top edge of the pool, and it expands to eliminate folds and wrinkles when filled with water.

Decks, ladders, and fencing are also made of extruded or sheet aluminum, and they are designed to be installed with the pool (Fig. 1-14). Of course, wood, plastic, and other materials can also be used, depending on budget and cosmetic preferences. Many excellent manufacturers

FIGURE 1-12 Embossed-metal side panels.
Northwest Wholesale.

FIGURE 1-13 Pool liner that resembles tile.

build a wide variety of aluminum pools, but some good examples are Delair Group's "Esther Williams" and "Johnny Weismuller" lines. Check the reference section at the back of this book for websites listing other fine makes and models.

STEEL

Some above-ground pools are made from galvanized or stainless steel components, similar to their aluminum counterparts and also finished with a vinyl liner (Fig. 1-15). The steel is galvanized and coated for rust prevention, while some edge, seat, or top beam components are covered with plastic or resin to keep them from being too hot to touch in the summer sun. The obvious advantage of steel over aluminum is that it has greater strength and generally lower cost, but the disadvantage is that it has greater weight and less resistance to rust.

That last problem is solved by at least one company, Artesian Pools, that markets an above-ground pool using all stainless steel components (Fig. 1-16). The stainless steel pool is coated with a liquid resin to prevent staining and allow a variety of colors to be added. It is also topped off with high-density resin seats, rims, and uprights. This manufacturer, like several others, also covers the interior surfaces of the stainless steel wall panels with an epoxy barrier that inhibits condensation between the wall and the vinyl liner. Such trapped moisture could result in mold or mildew. Artesian also offers decks, rails, and step units made with the same materials and coatings or extruded aluminum. Artesian is also one of the few companies that provides a lifetime warranty that is not pro-rated.

While we're on the topic, what is the significance of a "pro-rated" warranty? Many above-ground pool or vinyl liner manufacturers offer warranties on their products that vary from 10 to 50 years. These warranties are pro-rated, based on the number of years you have actually used the product and the number of years remaining on the warranty. The company will gladly replace the pool, liner, or component, giving a credit for the "unused" portion of the warranty period as applied to the cost of the replacement unit. The problem is that this pro-rated price for a replacement part is often greater than the price of a new one if purchased at one of the many discount supply stores or online marketers. The multidecade warranty sounds good when you're buying the pool or liner, but in reality it is only a way for ensuring the manufacturer's repeat business. When buying an above-ground pool or liner, ask the retailer for a realistic estimate of the life of the product and adjust your thinking accordingly. If you do call on the manufacturer for replacements, shop around and compare the cost of new pools and components first.

Like their aluminum counterparts, steel pools are manufactured by numerous outstanding companies, including Cantar, Cornelius Pools of Canada, Doughboy, and Ovation Pools, the last-named calling their products "raised pools." Like most high-quality metal above-ground pools, these companies use a special interior coating to inhibit moisture condensation and design their frame and base components to snap together for easy assembly. Edge and uprights are covered in various plastics and are designed to look like Greek or Roman columns (Fig. 1-17A and B). Cornelius and Doughboy use a unique steel buttress system

FIGURE 1-14 Typical deck, ladder, and fencing. *Delair Group, LLC.*

FIGURE 1-15 Typical steel-wall pool. *Cantar/Polyair Corp.*

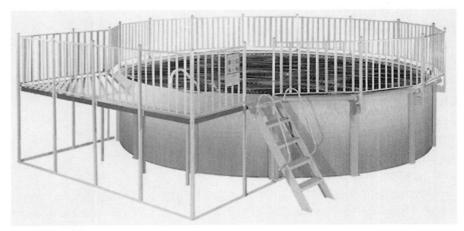

FIGURE 1-16 Stainless steel pool, deck, rails, and ladder. *Artesian Pools.*

that is recessed into the ground (Fig. 1-17C) and is designed to consume the least amount of space outside of the pool walls themselves.

Strong Industries, maker of everything plastic from caskets to pet baths, also manufactures a "granite resin" above-ground pool, shaped to look like stone columns, constructed over steel. The point here is that most above-ground pools that use metal components mix them with increasingly creative combinations of other materials for more attractive, utilitarian finished products.

PLASTIC

Plastic is the general term for the rigid, nonmetallic materials used to make some above-ground pools. The architecture is similar to metal pools, including a frame system, wall panels, and a vinyl liner. Plastics work well because they are completely rustproof and are extremely light yet sturdy. The drawback to using plastics is that they become brittle over time, and plastic components are more easily cracked or chipped than metal components.

Injection molding is used for many of these types of pools. Others use a cast resin process, which allows distinctive patterns to be formed to make uprights, seats, rails, and other features. Because plastic is not as strong as metals, these pools are typically round, rather than oval, so that the pressure of the water is more evenly distributed.

Sharkline makes an excellent line of above-ground pools using components made entirely of resin. Their unique patented design

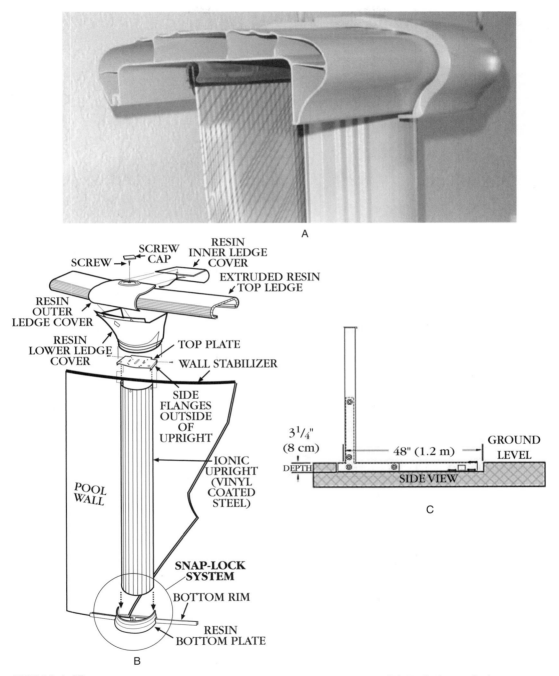

FIGURE 1-17 (A) Cutaway view of pool wall, column, and plastic cap. (B) Typical strut/column construction. (C) Narrow buttress system. *B: Zodiac American Pools, Inc. C: Cornelius Pools, LLC.*

FIGURE 1-18 Wood above-ground pool. *Technobois.*

allows the top rail to pivot to the "up" position for easy assembly, eliminating the need for metal fasteners and tools.

WOOD

Have you ever enjoyed a pleasant evening in a redwood hot tub? Well, at least two companies make above-ground pools out of southern yellow pine or western cedar (Fig. 1-18). Technobois in Canada makes pools that resemble their metal counterparts but have few or no braces because the 2-inch-wide by 6-inch-thick (5- by 15.5-centimeter) wood panel components do all the work. A 51-millimeter-thick vinyl liner is used to complete the pool, and decks or stair units are designed as attractive additions. Plastic columns and edge caps provide finished edges and simply snap together around the wood panels. These pools come in the same dimensions as other above-ground varieties and have the advantages of beauty and greater insulating properties than metal or plastic. Of course, they are less portable than pools made of lighter materials and require a bit more maintenance. That said, I have customers with hot tubs that are 40 years old and still in great shape, so durability is not an issue if properly maintained. Crestwood Pools in New York is another fine manufacturer of wood pools.

Onground

Onground pools are generally less expensive and more portable than above-ground pools. They are manufactured most commonly as round pools, up to 24 feet (7.3 meters) in diameter and the same depth as above-ground styles, although without the option of a deep end. Onground pools are not designed for use with an elaborate deck, fence, or stair system, but typically they have an A-frame ladder for easy access. Perhaps the most important feature of the onground pool is the ease of setup, teardown, and maintenance. This makes the onground pool an excellent choice for families on a tight budget and for those who prefer to store their pool during cold winter months, setting it up again each spring.

INFLATABLE

The most familiar onground pool is the inflatable variety. No longer just "kiddy" pools, inflatables are an inexpensive and portable solution to the swimming needs of many, and, as illustrated by the two different pools in Fig. 1-6, they come in the same dimensions as other above-ground pools. Inflatables are made of rugged PVC-vinyl material, and they require little ground preparation before setup. They are also extremely portable and easy to store.

SOFT-SIDED/FRAMED

The soft-sided/framed pool (Fig. 1-19) is the fastest-growing category of onground or above-ground pool because of its affordability, utility, portability, and ease of maintenance. They are also made with a wide variety of accessories in all sizes, shapes, and colors. Doughboy, Intex, Laguna, Splash SuperPools, Splash-A-Round, and Tuff Pools are just a few of the excellent manufacturers of soft-sided/framed pools.

The frames of these pools are typically tubular galvanized and coated steel (especially oval or rectangular shapes) or aluminum (for round pools, where water pressure is more evenly distributed). The laminated

FIGURE 1-19 Typical onground soft-sided/framed pool. *Splash SuperPools, LLC.*

FIGURE 1-20 Soft-sided/framed spa.

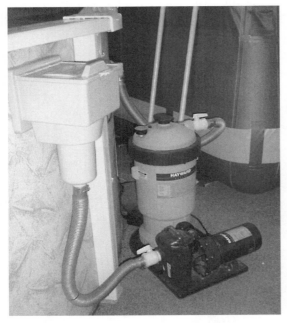

FIGURE 1-21 Typical equipment skidpack.

PVC-vinyl liner is usually thicker than above-ground styles, up to 40 millimeters. For added protection from puncture or abrasion, most manufacturers provide a ground cloth to be laid out before the pool is set up. Liners are also sold in various colors and decorative designs. Frame components simply snap together, and even the largest soft-sided/framed pool can be set up in an hour by two people. Most are sold as complete kits, with filtration equipment, plumbing, ladder, maintenance kit, cover, and an instructional video.

One other nice feature of these pools is that because of their simplicity and strength, they can easily be made into portable hot tubs, with added depth for standing or seated enjoyment (Fig. 1-20).

FRAMELESS

OK, so we figured out how to build a pool without walls, but could we build one without a frame? Amazingly, the answer is yes, and the result is strong enough to withstand the impact of a truck (look back at Fig. 1-7B).

Sofpool, Intex, and Muskin are a few of the companies that make these remarkable pools. Frameless pools are made from a woven fabric reinforced vinyl, just like NASCAR racing tires, that is typically more than 60 millimeters thick. You simply add water to the pool, which expands as it fills. An inflatable collar is added to the top edge to act as a seat, but is not part of the structure. When emptied, even the largest frameless pools fold up into a large duffel bag. Oval or rectangular versions have a few braces added along the length of the pool, but the round versions require no such stabilization. These pools are the ultimate in portability and affordability, but they are designed to utilize standard filters, heaters, automatic cleaners, ladders, and any other accessories you would find on other above-ground or onground pools. Equipment for onground pools is often sold on a single platform, called a *skid-pack*, for ease of installation (Fig. 1-21).

QUICK START GUIDE: PURCHASING THE RIGHT ABOVE-GROUND OR ONGROUND POOL—AS EASY AS 1-2-3

Here is a simple checklist of three key features that will direct you to the right pool choice, listed in order of priority for most consumers:

1. Size
   - Depth: If you need a pool large and deep enough to dive or jump in, buy an inground pool. There is no onground or above-ground pool designed for these activities, and, in fact, every manufacturer strongly warns against using their products for these purposes. It just isn't safe, even with the optional deeper swimming end that is designed into some above-ground models.

   - Family use: A good rule of thumb is to allow 15 square feet (1.4 square meters) of water surface for each bather. A 10- by 20-foot (3- by 6-meter) pool has 200 square feet (18 square meters) of surface, or enough room for up to 13 bathers.

   - Swimming laps: Length is more important than bather load, so you may want to choose a rectangular pool of sufficient length for a good workout.

   - Yard space: Regardless of the intended use, you are limited by the size of your yard, especially the area that is mostly flat. You also need to allow at least 3 feet (about 1 meter) on all sides of the pool (see installation guidelines below). Measure twice before buying the pool!

   - Size matters: There's nothing worse than an undersized pool. You may have a small family, but watch how fast it grows when you have a pool. Allow for the largest party or gathering you are likely to have when selecting a pool. That said, bigger means more of everything—initial price, cost of replacement liners, chemicals, cost of fuel to heat the pool, cleaning time, larger covers, more expensive equipment, and probably larger, more costly decks and ladders. Think about these extras when choosing

(Continued)

the pool size that is truly right for you. Unlike an inground pool, you can always start with a smaller above-ground pool and trade up if needed. Above-ground pools can be sold, moved, and set up easily in someone else's backyard.

2. Price

- A good-quality, 20- by 35-foot (6- by 10-meter), metal-sided, above-ground pool will cost around $4000, including standard filtration equipment, ladder, and sand or other materials to prepare the ground. Professional installation costs up to an additional $1000, with some installers charging $25 per inch (per 2.5 centimeters) to level the ground beyond the first 3 inches (7.6 centimeters). Smaller versions, easier to manage for the do-it-yourself owner, can cost as little as $300 for a 15-foot (4.6-meter) diameter (round) shallow model, including a simple filter/pump circulation unit. A good-quality soft-sided/framed pool of 10 by 20 feet (3 by 6 meters) will cost around $3000, requires little or no ground preparation or materials, and is easily installed by the consumer. A basic equipment package and ladder are typically included.

- A frameless round onground pool of 16 feet (5 meters) in diameter will cost under $1500 including basic equipment and ladder. A 10-foot-diameter shallow model with a small circulation pump can cost less than $200 at major mass-market retailers. Of course, these are the easiest to set up for the average consumer and require no other preparation.

- Other price factors: (1) The larger the pool, the larger the pump you need. You will also need a main drain and additional return lines. These can add a few hundred dollars to any choice of pool. (2) Decks and fencing are valuable additions for comfort, safety, appearance, and ease of pool maintenance. Standard deck/fence kits will cost from $500 to $2000, depending on size and materials. The price of custom wood decks and stair units will vary greatly, depending on size, material (redwood is far more costly than pine, but it lasts longer), and complexity.

3. Setup and teardown

- Year-round pools: If you don't want or need to disassemble (or winterize) your pool each year, a sturdy, long-lasting. metal-sided above-ground pool may be the best investment.

- Annual closure and mobility: If you need to tear down the pool each winter or plan to move, you may prefer the superior portability of the soft-sided onground pools.

- Landscaping: You can landscape around your above-ground pool, but do not set the pool in the ground and backfill. Above-ground pools are not designed to be buried, and the force of earth against the exterior wall, pushing inward, can cause component deterioration and failure.

FIGURE 1-22 **Typical onground frameless pool.** *Sofpool, LLC.*

Installing an Above-Ground Pool

Frameless onground pools (Fig. 1-22) are so simple to set up that almost anyone can figure them out, even without the instructions in the owner's manual, so this section will follow the process of installing a typical above-ground metal pool and a soft-sided/framed onground pool. These guidelines will not make you a pool installation expert because each make of pool sets up in slightly different ways, but if you follow these steps, you will be able to decide if the do-it-yourself approach is right for you. Moreover, you may be dealing with a pool already in use, so it is worth understanding the overview of how that pool came into being. This knowledge will allow you to estimate the maintenance costs, future potential repair expenses, or the problems involved in upgrading it.

TOOLS OF THE TRADE: TYPICAL ABOVE-GROUND POOL INSTALLATION (ALL TYPES)

- 25-foot tape measure
- 8 large wooden stakes and a ball of string
  - OR a 2- by 4-inch (5- by 10-centimeter) board that is as long as the radius of the pool
- Hanging level (the type that hangs from string)
  - OR a standard carpenter's level that lies along the edge of the board
- 1 center stake and length of string equal to pool's radius.
- Spray paint (bright color)
- Shovel
- Tamping tool
- Masonry sand (to fill small potholes)
- 12-inch-square (30-centimeter) patio stones, one for each base plate
- Hammer
- Screwdriver set
- Pliers
- Box wrench set
- Nut driver set or socket wrench set
- Gloves
- Clothespins and/or wide-mouth paper clips
- Flex PVC pipe, glue, primer
- Teflon tape or pipe dope (if required)
- Fine-tooth hacksaw
- Razor knife
- Awl (or large nail)
- Duct tape
- Silicone sealant
- Vacuum cleaner
- Bottom liner material and seam tape

- Shears for cutting the bottom liner
- Prefabricated cove material
- A-frame ladder
- Garden hose attached to your water supply
- Towel (to wrap the end of the hose when filling)
- Sanitizer

Typical Onground (Soft-Sided/Framed) Pool Installation

RATING: MODERATE (MINIMUM TWO WORKERS)

TIME: 2 HOURS MINIMUM (LONGER FOR LARGER POOLS)
PLUS THE TIME TO LEVEL YOUR GROUND

1. **Choose the Location** Take time to select the best location for your pool, which will save you hours of maintenance or repair problems later, using a few common-sense criteria:

   ■ Adhere to local codes and standards. Most jurisdictions do not require permits for portable pools, but if in doubt, check with your local building department or a building contractor.

   ■ Avoid areas with overhanging wires or rooftops. People often ask that the pool be installed with part of a patio or garage roof overhanging to create a shaded area in the pool, but this is almost never a good idea because dirt and leaves from the roof will wash directly into the pool every time it rains.

   ■ Avoid trees. Branches overhead will make pool maintenance a chore, and heavy debris can clog plumbing, causing damage to pool equipment. Tree roots also have a way of punching up through pool liners or, at a minimum, creating uneven surfaces that are hard on your feet and your automatic pool cleaner. In addition, the uneven surfaces form great crevices in which algae can grow.

   ■ Avoid areas with buried pipes, septic tanks, sprinklers, or cables that might need servicing in the future or that might be damaged by pressure from the weight of the pool.

QUICK START GUIDE: INSTALLING AN ONGROUND SOFT-SIDED/FRAMED POOL

1. Prepare the site
   - Level the ground. Cut down high spots—don't fill in low spots.
   - Lay a protective sheet of vinyl or install a bottom liner.

2. Lay out the pool
   - Read all installation instructions.
   - Take the framework, hardware, and plumbing out of the boxes, and lay them out in an area adjacent to the pool site.
   - Lay out the pool itself and unfold it onto the prepared site.
   - Identify the prepunched holes for plumbing connections; situate that end closest to the source of electricity (where you will assemble the equipment).

3. Assemble the pool
   - Insert the top rail components into the pool sleeve(s) and connect them (typically threaded, snap-together, or locking pins). Start with one of the longer sides (if your pool is not round).
   - Raise the pool. Lift one side of the top rail; each upright will move toward the base of the pool, where straps attach the bottom of the upright securely. Work around the pool until all uprights have been firmly attached to their support straps.

4. Inspect your work
   - Walk around the pool, pulling out any wrinkles.
   - Check that each upright is still fully inserted into the top rail.

5. Install the equipment
   - Set the skidpack on level ground.
   - Connect the plumbing to the pool and skidpack as described in your owner's manual. Tighten all connections BY HAND ONLY. Add the skimmer (if so equipped).
   - Have a waterproof electrical outlet, with a ground fault circuit interrupter, available. Do not plug in the skidpack yet.

6. Fill the pool
   - Fill to within 3 inches (8 centimeters) of the top or until the intake port (or skimmer, if so equipped) is completely flooded.

7. Start the circulation equipment
   - Open all valves and plug in the skidpack to turn on the circulation pump.

- **Check that water circulates through the system and back out the discharge port.**
- **Bleed air from the system by opening the bleeder valve on the top of the filter.**
- **When the water is circulating properly, add the first dose of sanitizer.**

8. **Enjoy your new pool**
   - **No diving or jumping!**
   - **Add decks, ladders, and rails.**

- Consider the distance from the pool site to the source of water (remember, you need to refill your pool every week to replace evaporated water) and electricity for pool equipment.

- Consider where your pool will drain. Someday, either due to leaks or maintenance chores, you will drain all or part of your pool water. Make sure you're in an area that can accommodate a flood, or consider adding the plumbing necessary to drain your pool through the pump and out through a backwash hose (see Chap. 5).

- Allow adequate side clearance. The most obvious location criterion is the space needed for the pool you have selected, but surprisingly, most people assume their yard is adequate and fail to measure. If your pool uses A-frame bracing, it will require more space around the perimeter than vertically braced or frameless pools, but plan for 4 feet (1.2 meters) around any pool. That leaves enough room not only for plumbing access and equipment but also for workers when installing the pool itself. You can later add landscaping that is closer to the pool walls themselves, but leave plenty of room around the intended perimeter of the pool for the installation procedures.

- Keep the pool at least 6 feet (2 meters) away from steep slopes, which can result in erosion of the base area by heavy rain runoff.

- Consider the prevailing wind. Try to avoid being downwind of trees or other potential sources of debris. The wind can assist your water circulation, so think of where that will be located and plan to align your pool with the suction port (or skimmer, if so equipped) downwind.

- Never put your pool on a wooden deck. Decks are just not designed to carry that much weight concentrated in so small an area. Moreover, wood and fasteners have a way of separating or splintering, causing damage to the pool.

- Think ahead: Will you want to add a deck, slide, landscaping, or fencing later? What about a heater? If so, you'll want to allow more space around the pool. Also think about where the gas will come from to fuel a heater. A solar heater takes up even more land area, but it warms your pool for zero fuel cost (see Chap. 3). You will also want to take the orientation of these extras into consideration: do you want the deck between your house and the pool or on the far side of it? Should the ladder be visible from the house to monitor who goes in the pool?

2. **Prepare the Site** Generally speaking, site preparation is a matter of making the ground level and free of abrasive materials. Warranties can be voided if these conditions are not met in accordance with the manufacturer's requirements because rough surfaces can damage the fabric or liner, and setting a pool on a surface that is not level will place excessive stress on the low side(s).

First, measure and mark out the area your pool will occupy. Next, use your shovel to remove the grass, stones, and any other abrasive materials. Then level the ground to within 1 inch (2.5 centimeters) of level as measured at opposite sides of the pool. When leveling, you should remove high spots rather than fill in the low spots. This is because fill material will compact under the weight of water in the pool, causing low spots to develop all over again.

There are two simple tools that can be used for determining if the surface is level. The best is a 2- by 4-inch (5- by 10-centimeter) board of sufficient length to span the radius of the pool. If your pool is oval or oblong, work in sections (refer to the capacity calculations in the beginning of this chapter). Place one end of the board where the center of the pool will be located and the other end in any direction (you'll cover all 360 degrees of the compass before you finish). Set the carpenter's level along the edge of the board to determine if the ground is level, scraping down high spots with the shovel and rechecking until you're sure it's level. Keeping one end of the board in that center location, rotate the board around a few feet and check the level again. Make adjustments and repeat the process until you have checked the entire 360 degree circle. One benefit of the board system of leveling is that you can use it to scrape the ground as you rotate it through the circle (assuming your helper is keeping it anchored firmly at the center point), removing minor high spots in the process.

The second method of leveling is to use stakes and string. This method is preferred if the length or radius of your pool is longer than a typical board. Figure 1-23A shows how to use stakes, string, and a line level to accomplish the same goal as described in the previous method. Here too you will work in a circle (or series of circles for larger pools). The string method is much easier for oval pools because you can run the string the entire length of the pool and repeat the process across the width at several points. The most critical element of using the string method to level a site is to ensure that the string is held taut between the stakes at all times. Sagging string means inaccurate results.

When you have completed the ground preparation, be sure to remove any stakes, wood, or other debris you may have left inside the area. I have been called to repair pool leaks and discovered a punctured pool bottom because someone forgot this simple step!

One more note about ground preparation. Don't just install your pool on top of the grass, even if it's already level. The grass will die in time, but before it does, it can punch through the liner. Moreover, grass will compact unevenly under the weight of the water, leaving wrinkles in the liner and uncomfortable surfaces under your feet.

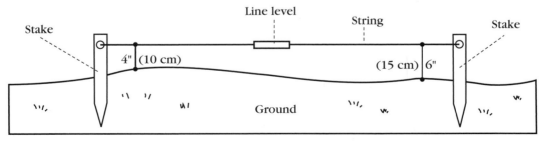

FIGURE 1-23A Level the ground before installing the pool.

Dead grass will also decompose, creating pockets of gases and bubbles underneath the pool. Consider the additional investment in a bottom liner, which is a sheet of heavy vinyl or foam that you lay on the ground before assembling the pool. It will help protect your pool bottom from punctures and create a smoother surface for bare feet. More description of bottom liners is provided in the section on installing a metal-walled above-ground pool (see Fig. 1-34), because they are more commonly used with those models.

3. **Lay Out the Pool** Read the installation instructions cover to cover before beginning. The investment of a few minutes at this point will pay big dividends during the installation process. Start by taking all framework, hardware, and plumbing out of the boxes and lay them out in an area adjacent to the pool site. Lay out the pool itself and unfold it onto the prepared site. Don't drag the pool around because even small abrasions can weaken the fabric over time. Have your helpers lift the pool with you as you unfold or reposition it. Identify the part of the pool with the prepunched holes for the plumbing connections, and situate that end closest to the source of electricity (where you will assemble the equipment). Remember that the goal is to position the suction port or skimmer downwind so that debris is naturally carried into it.

   Lay the framework components around the pool before assembling them to be sure you understand how they all fit together and you can see that everything is positioned correctly. Some manufacturers number the parts of the top rail so there are no mistakes: 1 attaches to 2, 2 attaches to 3, and so on. If your top rail parts are numbered, lay them out in order.

4. **Assemble the Pool** Typically, the upper edge of the pool will be fitted with sleeves that accommodate the top rail. Insert the rail components into the sleeve(s) and connect them as instructed by the owner's manual. Rail pieces are typically threaded, snap together, or are connected with locking pins. Start with one of the longer sides (if your pool is not round). As you connect sections of the rail together and slide them into the sleeve, be sure to align the holes in the rails with the holes in the sleeves that will later accommodate the uprights. To get a sense of what the assembly will look like as you make progress, take a peek ahead to Fig. 1-23F. Try to smooth out all wrinkles as you work.

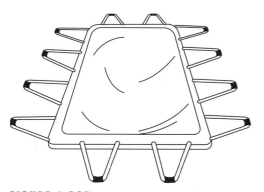

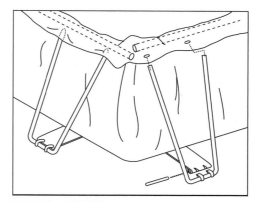

FIGURE 1-23B Lay out the pool components.

FIGURE 1-23C Setup of typical soft-sided/framed onground pool.

When the top rail has been completed, assemble the uprights (your model might call them *legs*, *buttresses*, *U braces*, or *vertical tubes*) and insert them into the top rail, but do not attach them to the fabric "feet" of the pool just yet. Some uprights are supplied as one-piece units, while others require some assembly to create U-shaped components. If your pool uses vertical uprights (rather than the slanted A-frame style), assembly will be similar, but the base of the upright will fit into the base of the pool in sleeves similar to the one provided for the top rail.

Uprights typically do not lock into the top rail but slip-fit securely into the hole provided. The weight of the water will adequately lock the uprights into the top rail. When you complete this process, your installation should look something like Fig. 1-23B, all lying flat on the ground.

Now it's time to raise the pool. Have one or more helpers stand inside the pool (shoes off to avoid scratching the liner!) and lift one side of the top rail to the normal height. As the pool is raised, the base of each upright will move toward the base of the pool, where fabric straps are attached to receive them. These *upright straps* (also called *support straps* or *buttress straps*) are designed to accommodate hardware that attaches the bottom of the upright securely to the strap as shown in Fig. 1-23C. Methods of attachment differ, but all will involve some kind of self-explanatory clip, ring, or locking pin. (Note that Fig. 1-23C shows how the top rail inserts, but this process should be complete by this point.) Work

FIGURE 1-23D   Buttress block.

around the pool, smoothing out all wrinkles wherever you can, until all uprights have been firmly attached to their support straps and the pool shell is fully assembled.

When the pool is filled, each upright will be supporting as much as a ton (almost 1000 kilos) of water. Figure 1-23D shows typical *buttress blocks,* which are heavy rubber or plastic feet that distribute that weight more evenly, especially on soft ground. You may not need buttress blocks if your pool is assembled on concrete, but they're a good idea otherwise. Buttress blocks will be designed for your particular pool and will attach as part of the process that connects the upright to the support strap.

5. **Inspect Your Work** With everyone out of the pool, perform one last check of your assembly. Walk around the pool, pulling out any wrinkles and checking that each upright is still fully inserted into the top rail (Fig. 1-23E). An upright that has popped out during assembly of the other side can look normal at this point, but it will cause the pool to collapse when it is being filled with water. DO NOT SKIP THIS CRUCIAL STEP!

6. **Install the Equipment** Most onground pools are sold with a package of pump/motor and filter, along with all necessary plumbing and electrical connections. Figure 1-23F shows a typical installation.

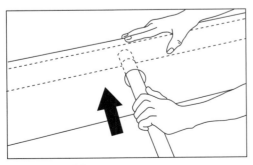

FIGURE 1-23E   Securing uprights to top rail.

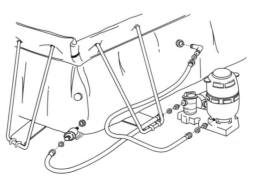

FIGURE 1-23F   Typical equipment skidpack.

Set up the equipment package, called a *skidpack,* on level ground away from sprinklers and close enough to a source of electricity for the cord to reach it. If your pool is close to a deck or other hard surface, it's a good idea to install it there, although you may want to bolt it down since it will more easily vibrate under these conditions.

Connect the plumbing to the pool and equipment as described in your owner's manual. Most pools clearly mark *intake* and *discharge ports,* and the plumbing is designed to be simply threaded in place with no Teflon tape or pipe dope required. (See Chap. 2 for basic plumbing descriptions of these products if your installation does require them.) Tighten all connections BY HAND ONLY. It's better to have a leak and tighten the connections further than to overtighten them in the first place and crack the fittings (which is easier to do than you might think, especially as the plastic components age and become more brittle).

As with the other parts of installing your pool, you must read the owner's manual fully before adding water or turning on the pump. Although most systems are designed for "plug-and-play" operation, some manufacturers might pack spare parts inside a filter housing or strainer pot, and some may ship components that are not fully assembled. Trust me, the time it takes to read through the manual(s) will make your installation (and use) much easier and safer.

Most skidpacks have long cords that simply plug into household current, but make sure cords will not trip anyone or be sliced by the lawnmower. Also ensure that cords are plugged into waterproof outlets serviced by a ground fault circuit interrupter (see Chap. 7 for more details). If you plug the unit into a timer, which is then plugged into the wall outlet, be sure to buy the weatherproof timer unit.

7. **Fill the Pool** With everything assembled, it's time to fill the pool, but don't toss in the garden hose and walk away. First, loosely wrap a towel over the end of the hose so the metal fitting doesn't whip back and forth on the bottom, causing scratches. Tape or tie the towel securely and lay the hose into the pool. You can also just hold the hose or secure it to the top rail, but you'd be amazed how much water evaporates when you fill a pool with that "cascade" method, especially on a hot day.

As the pool fills, watch for leaks or wrinkles that develop. Wrinkles can be smoothed out as the pool fills, but probably not when it's full. Make adjustments to the pool, top rail, and support straps as needed, especially after the first inch (2.5 centimeters) of water has been added. Fill the pool to within 3 inches (8 centimeters) of the top or until the intake port (or skimmer) is completely flooded.

While you monitor the filling of your pool, you can use the time to assemble any ladders or other ancillary equipment you may have purchased. These components are usually very simple to assemble, but as always, follow the owner's manual.

8. **Start the Circulation Equipment** First, a word of caution. Any pool pump generates considerable suction, but the systems created for onground and above-ground pools can be uniquely forceful. This is because the equipment is usually located close to the pool and the suction is concentrated at only one suction port, much like a spa or hot tub. This suction can easily pull in hair, hands, clothing, or skin. ALWAYS turn off the equipment before adjusting or servicing it.

If the water has now flooded the intake port (or skimmer), make sure all valves in the plumbing are in the *open* position (usually the valve handle is parallel to the pipe for open, across the pipe for closed) and turn on the circulation pump. Check that water is circulating through the system and back out of the discharge port (it may take a minute or two to purge the air in the components before you see currents of water discharging). Bleed air from the system by opening the bleeder valve on the top of the filter, then closing it when water spits out instead of just air. Additional tips about operating your pump, motor, and filter are found in other chapters of this book.

When the pool is full and the water circulating properly, add the first dose of sanitizer to achieve an appropriate level (see Chap. 8) for the product you have chosen. Be sure the water is circulating freely around the pool so the sanitizer distributes evenly.

9. **Enjoy Your New Pool** It might seem unnecessary to describe the use of your pool, but the reason for this section is safety. Not a

FIGURE 1-23G Typical prefabricated deck unit. *Splash SuperPools, LLC.*

single above-ground or onground pool that I have ever seen is designed for diving or jumping. In reviewing the owner's manuals from dozens of manufacturers, every one of them warns against using their pools for diving or jumping. The fact is that even with an optional deep end, above-ground and onground pools are just not large enough or deep enough for this purpose. Having said this, there are large commercial versions of above-ground pools with decks and diving boards, virtually indistinguishable from inground pools. These are specially designed and usually installed in public settings where a lifeguard is also standard equipment. Even if you install a slide, ladder, deck, or other "entry" equipment (Fig. 1-23G) on your pool, I urge you to prevent anyone from diving or jumping into it. It's just not safe!

TRICKS OF THE TRADE: INSTALLING YOUR ABOVE-GROUND POOL

- To get a better understanding of space requirements, make a pattern out of newspaper that you tape together and lay out in the same dimensions as the proposed pool. This template can be easily moved from one location to another or angled in various ways to test out just what it might look like in various parts of the yard. When making the template, include the 4-foot (1.2-meter) clearance you need around the perimeter of the pool itself, so you're creating an accurate picture of the final installation.

- The sturdy roots of nut grass, "bamboo" grass, and Bermuda grass have a way of growing right through plastic liners, even if you think you have removed all the plant material on your site. If your yard has one of these hearty types of grass, treat the area with a nonpetroleum herbicide after scraping the area clean, and be sure the clearance area around the pool perimeter is also free of these grasses.

- On hot days, the rails may stick to the fabric when inserting them into the vinyl sleeves of onground pools. If they do, wet the rail before inserting it, but in any case, never force a rail into the sleeve. You may be jamming against a fold or other part of the pool fabric, and any time that metal competes with vinyl, the metal always wins. Don't use soap to lubricate balky rails—if it gets in the pool, it will create suds when the circulation system is turned on.

- If the plumbing connections leak when you fill the pool, and further tightening hasn't helped, check to see if you missed installing an O-ring or gasket at any point. Reread the owner's manual and look around the ground for errant parts.

- Most pools provide warning labels that are designed to be attached to ladders, equipment, suction ports (skimmer and main drain), and decks. USE THESE STICKERS. You will have guests someday who may not understand that an above-ground pool is not the same as the others they've used, and safety should always be the first consideration.

- Don't try to install your pool on a windy day. Liners, metal walls, and other large components are tough enough to handle without fighting the wind. Remember too that you'll be exposing bare soil, which can blow into your eyes as you work on windy days.

QUICK START GUIDE: INSTALLING AN ABOVE-GROUND (RIGID-SIDED) POOL

1. **Prepare the site**
   - Trace the outline of the pool walls with a stake, string, and a can of spray paint (Fig. 1-24).
   - Level the ground. Cut down high spots—don't fill in low spots.
   - Excavate optional deep end (Fig. 1-25) and main drain (and plumbing trench) as needed.
   - Lay out the bottom rail to determine the location of patio stones (one for each base plate); set the stones level with surrounding ground.

2. **Lay out the pool**
   - Read all installation instructions.
   - Lay out all pool components adjacent to your work area.
     - Keep metal components out of direct sun!
   - Set a base plate on each patio stone, and slide the base rail into the base plates all around the perimeter, connecting the rail into one big circle (Figs. 1-26 and 1-27).
   - Make sure you have a truly round shape at this point and adjust it as needed.
   - Check the level of the bottom rail by laying your carpenter's level along the rail between each plate.

3. **Assemble the pool**
   - Put the roll of sheet metal inside the perimeter of the pool area and stand it up on cardboard (Fig. 1-28).
     - Be sure the skimmer opening is on top!
     - Place an A-frame ladder inside the pool perimeter so you can get back out when the wall is fully assembled.
   - Uncoil the pool wall and walk it around the interior of the perimeter, setting it into the bottom rail (Fig. 1-29).
     - Be sure that the skimmer and discharge openings are not aligned with any of the base plates.
     - As you work, apply the slotted tubes (Fig. 1-30) to the top of the pool wall (and the special retainer rail if you are using a beaded liner).
   - Bolt the wall ends together (Fig. 1-31)
     - Put the round head of the bolt on the inside of the pool and the nut on the outside.

(Continued)

- Apply three layers of duct tape over the heads of the bolts.
- Install the uprights to the bottom plates (Fig. 1-32).
  - Only secure uprights to the base plates at this point.

4. Inspect your work
   - Measure the top of the pool walls across several points to be sure you have a truly round or squared shape (Fig. 1-33).
   - Use your carpenter's level again to ensure that the entire structure is level.

5. Complete the bottom preparation
   - Lay the main drain (and pipe to the skidpack) in place if your pool is so equipped.
   - Secure this assembly in place by filling the gaps with dirt and compacting it firmly by hand, then covering it with masonry sand. Make sure the result is level with the surrounding ground.
     - Remove the main drain cover, retaining ring, and gasket, and set these aside until the liner has been installed.
     - Raise the open end of the main drain pipe up and tape it to the top of the pool wall.
   - Cover the ground with clean, fine masonry sand to cover any minor potholes.
   - Install a protective foam or vinyl bottom liner (Fig. 1-34).
   - Install the cove around the pool base interior (Figs. 1-35 and 1-36).

6. Hang the liner
   - Install the pool liner (Fig. 1-37).
     - Lay it out in the sun to relax creases and make the vinyl more supple.
     - Punch out openings for the skimmer and return lines.
     - Deploy the liner around the inside of the pool, hanging the edges over the side of the pool walls or inserting the bead into the hook (Fig. 1-38). The round seam should look like a line drawn evenly around the cove. The cross seams should be parallel and not running over the skimmer or return outlet cutouts in the pool walls. The material should lie smooth and snug against the pool wall, the cove, and the bottom of the pool. Do not trim off the excess liner.
     - Secure by attaching the inner combing and plastic coping along the top edge of the pool wall.
   - Complete the main drain installation (Fig. 1-39) by assembling the retainer ring and gasket to the main drain unit. Then carefully cut the liner away from the interior of the drain. Attach the main drain cover.

7. **Fill the pool**
   - **Start filling the pool and smooth out new wrinkles (Fig. 1-40) by removing a section of the coping and adjusting the liner up or down.**
     - **To assist with wrinkle removal, use a vacuum cleaner (Fig. 1-41).**
   - **Don't add more than 6 inches (15 centimeters) of water until the uprights have been fully attached.**
   - **Secure the uprights to the top of the pool wall, adding the top plates.**
   - **Secure the top rail all around the perimeter (Fig. 1-42).**
     - **As you work around the pool, you may need to push the top of each upright in toward the center of the pool to bring it into alignment. This will also help the bottom kick slightly outward, perfecting the circle of the pool.**
   - **Snap on any resin plastic decorative covers for the uprights or top rail (Fig. 1-43).**

8. **Install the equipment**
   - **BEFORE the water reaches the level of intake or discharge ports:**
     - **Install the return line discharge port.**
     - **Install the skimmer (Fig. 1-44). Cut the liner material out of the skimmer mouth and the interior of the vacuum port.**
     - **Plumb flex PVC pipe from the return line fitting to the equipment area. Connect the optional main drain pipe and the skimmer pipe to a three-port valve at the base of the pool wall.**
     - **Run another length of pipe from the three-port valve to the pump.**
   - **Finish filling the pool.**

9. **Start the circulation equipment**
   - **Plug the skidpack (Fig. 1-45) into a waterproof outlet equipped with a ground fault circuit interrupter when the water has flooded the skimmer.**
   - **Look for steady flow of water (after the air has been purged from the system) at the return line fitting.**
   - **Add the first dose of sanitizer in accordance with package directions.**

10. **Inspect for leaks**
    - **Look around the pool, plumbing, and equipment for leaks. The gaskets at the skimmer, main drain, and return line fitting will need to get wet and expand, so a slight leak may be normal at first, but it should seal within a few minutes.**

(Continued)

11. Add a ladder, slide, or deck
    - Add these options several days after you have installed your pool so you have had plenty of time to inspect it for leaks or defects.
    - Landscaping is also best added after you have used your pool for a few weeks.
12. Enjoy the pool
    - Never allow diving or jumping into an above-ground pool!
    - Most top rails are designed for bearing your weight as you get in or out of the pool and if you sit on the rail, but they are not designed for you to stand on.
    - Don't use the pool if the bottom is not clearly visible.

Typical Above-Ground (Rigid-Sided) Pool Installation

RATING: PRO (MINIMUM THREE WORKERS)

TIME: 4 HOURS MINIMUM (LONGER FOR LARGER POOLS)
PLUS THE TIME TO LEVEL YOUR GROUND

1. **Choose the Location** Follow the instructions described in the previous section. Several above-ground pool manufacturers recommend against installing your pool on concrete (out of concerns about abrasion of the liner), but I have seen many installed on smooth hard surfaces, including exhibit halls at pool and outdoor recreation shows. Indeed, one of the many benefits of onground and above-ground pools is their ability to be assembled almost anywhere. That said, if you are installing on a hard surface, take extra care in laying the bottom liner to protect your pool liner from abrasions and in any case follow manufacturer's guidelines to maintain your warranty. Asphalt or any other petroleum-based surface can eat right through the bottom liner and the pool liner, so it is unwise to assemble your pool on these surfaces.

2. **Prepare the Site** The general guidelines detailed in the previous section may be followed but with a few more preparations for above-ground installations.

    First, a precise delineation of the pool's footprint is needed for a successful installation. After clearing and leveling the ground as previously described, trace the outline of the pool walls with a stake, string, and a can of spray paint as shown in Fig. 1-24A. To do this, you will need to know the radius of your round pool or the length and the radius at each end of your oval pool. Drawing the outline of a rectangular pool is simply a matter of using stakes and string to map out the dimensions and spray paint to draw the lines (Fig. 1-24B).

As with onground pools, the key to success is being level to within 1 inch (25 millimeters) throughout the proposed site. Remember to carve down high spots rather than fill in the low ones. However, small potholes can be filled in with sand or soil as long as they are packed as hard as the surrounding surface. In these cases, don't trust your feet or slapping the fill material with the shovel. Get a heavy tamping tool and pound it down firmly. Pay special attention at this point to the actual perimeter of the pool, where the walls will be standing. This circumference must be absolutely level to within 1 inch (25 millimeters) all around.

If your pool will include an optional deep swimming end, you will need to excavate that section. Mark the dimensions using the template or measurements provided by the manufacturer. Figure 1-25 shows typical diagrams and reference charts. Select the size of your pool from the row at the top, then read the column below for the exact length of each dimension as depicted in the diagram. Since these measurements are meant as examples only, they have not been converted to metric.

If your pool includes a main drain, you will need to dig a trench from the center of the pool (or middle of the optional deep end) to the area where the equipment will be located. To adequately accommodate the pipe and to protect it

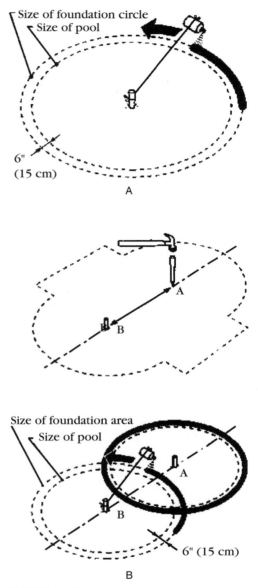

FIGURE 1-24 (A) Tracing the outline of a round pool. (B) Tracing the outline of an oval pool. *Cantar/Polyair Corp.*

Planning a Deep Swimming Area

An **Optional Special-Purpose Deep Swimming Area** is for underwater swimming only. **DO NOT DIVE OR JUMP!**

If an Optional Special-Purpose Deep Swimming Area is desired, excavate a deep end using the appropriate chart for your style pool. Our round, Com-Pac™ II Support oval pools with 20-or 25-mil expandable liners are designed for excavation.

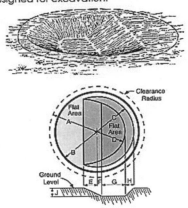

Round Dimensions

FIG.	12'	16'	18'	21'	24'	28'
A	7'	9'	10'	11'-6"	13'	15'
B	6'	8'	9'	10'-6"	12'	14'
C	5'-3"	7'-3"	8'-3"	9'-9"	11'-3"	13'-3"
D	3'-9"	4'-9"	5'-3"	6'-9"	8'-3"	10'-3"
E	3'	4'	5'-6"	5'-3"	6'	7'
F	1'-6"	3'-6"	3'-6"	3'-9"	3'	2'
G	2'-3"	1'-3"	1'-9"	3'	5'-3"	8'-3"
H	1'-6"	2'-6"	3'	3'	3'	3'
J	1'-6"	2'-6"	3'	3'	3'	3'

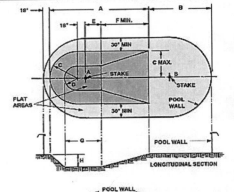

For pools with 20-or 25-mil expandable liners only. Using the center stake (stake "A"- round; stakes "A" & "B" - oval) as reference, select the dimensions for your pool size from the appropriate chart, then lay out the area and excavate. Remember: **DO NOT DIVE OR JUMP** into the pool. Excavations are intended to provide a deeper and wider swimming area only.

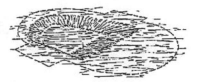

Com-Pac™ II Support Oval Dimensions

FIG.	20'X12'	24'X12'	28'X12'	24'X16'	28'X16'	32'X16'	34'X18'	38'X18'	41'X21'
A	12'-6"	14'-6"	17'-6"	15'-6"	17'-6"	19'-6"	21'-6"	23'-6"	25'-6"
B	6'-0"	8'-0"	9'-0"	7'-0"	9'-0"	11'-0"	11'-0"	13'-0"	14'-0"
C	3'-0"	3'-0"	3'-0"	5'-0"	5'-0"	5'-0"	6'-0"	6'-0"	7'-6"
D	1'-0"	1'-0"	1'-0"	2'-0"	2'-0"	2'-0"	3'-0"	3'-0"	4'-6"
E	2'-0"	4'-0"	7'-0"	0'-0"	2'-0"	4'-0"	5'-0"	7'-0"	7'-6"
F	6'-0"	6'-0"	6'-0"	9'-0"	9'-0"	9'-0"	9'-0"	9'-0"	9'-0"
G	4'-6"	6'-6"	9'-6"	3'-6"	5'-6"	7'-6"	9'-6"	11'-6"	13'-6"
H	2'-0"	2'-0"	2'-0"	3'-0"	3'-0"	3'-0"	3'-0"	3'-0"	3'-0"
I	2'-0"	2'-0"	2'-0"	4'-0"	4'-0"	4'-0"	6'-0"	6'-0"	9'-0"
J	2'-0"	2'-0"	2'-0"	3'-0"	3'-0"	3'-0"	3'-0"	3'-0"	3'-0"

IMPORTANT: If excavating your pool for an Optional Special-Purpose Deep Swimming Area, secure the excavated area against unauthorized, unsupervised, or unintentional entry. Also note that bad weather conditions (esp. rain) can cause the excavation to cave in. Plan accordingly before excavating.

FIGURE 1-25 Tracing the outline of the optional deep end. *Hoffinger Industries/Doughboy Pools.*

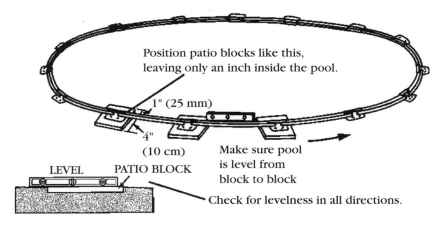

Position patio blocks like this,
leaving only an inch inside the pool.

1" (25 mm)

4"
(10 cm)

LEVEL PATIO BLOCK

Make sure pool
is level from
block to block

Check for levelness in all directions.

FIGURE 1-26 **Laying out patio stones, blocks, and bottom rail.** *Sharkline Pools.*

from damage when the weight of the water is on top of it, dig the trench about 1 foot (30 centimeters) deep and as wide as the pipe being used.

Above-ground pools benefit from supporting their uprights with concrete patio stones (some manufacturers recommend it while others require this additional preparation, especially on larger pools). Use 12-inch-square (30-centimeter) stones to support most uprights, and set multiple stones adjacent to each other as needed for larger buttress systems. The ground must be excavated carefully for each stone so that it is absolutely level with the surrounding surface (Fig. 1-26).

To determine the correct placement of the patio stones, you will need to lay out the base rails and base plates, and then use the spray paint to mark the exact location for each stone. As shown, place each stone so that about 1 inch (25 millimeters) lays on the inside of the bottom rail and the remainder lays outside the rail.

When you have completed the ground preparation and set all of the support stones in place, be sure to remove the center stake and any debris you may have left inside the perimeter of the pool outline.

3. **Lay Out the Pool** Read the installation instructions cover to cover before beginning this crucial step. At this point, you have already laid out the bottom rail and base plates to aid you in setting the support stones, but now lay out all of the other components adjacent to your work area. On hot days, keep metal side components and hardware out of direct sun

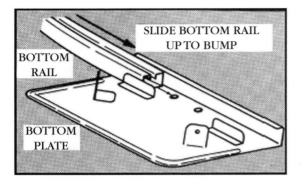

FIGURE 1-27 Base rail installation. *Sharkline Pools.*

or plan on wearing gloves as you work. While we're on the subject, wearing gloves during installation of a metal-sided pool is a good idea anyway—it's easy to scrape your hands or cut your skin on sharp edges.

Set a base plate on each of the patio stones you have set into the ground, and slide the base plate rail into the base plates, all around the perimeter, connecting the rail into one big circle (Fig. 1-26). Some manufacturers build in a stop bump into their base plate (Fig. 1-27) to let you know precisely where the end of each bottom rail section is meant to be. Others simply instruct you to leave a gap of 1/2 inch (13 millimeters). Either way, follow the instructions carefully so that the next steps of the installation are not hampered by incorrect assembly of these initial components.

Before proceeding, check to make sure you have a truly round shape at this point and adjust it as needed. Measure across the pool from opposite base plates, and be sure the measurement is the same each time. Now check the level of the bottom rail by laying your carpenter's level along the rail between each plate (Fig. 1-26). Make any necessary adjustments to the shape or level of your pool now to save a lot of time later. On oval pools, measure across the straight sides of the pool, from base plate to base plate, and use your stake and string again to determine that the round sides are equidistant. If it all measures up, you are now ready to assemble the pool itself.

4. **Assemble the Pool** Put the roll of sheet metal inside the perimeter of the pool area and stand it up on one end, resting it on the cardboard from the pool box (Fig. 1-28). Be sure you have the top side up, so that the skimmer opening will be on top as shown in Fig. 1-28. Believe it or not, it is a common mistake to have the top side down, and it is often not discovered until the entire pool has been set up and the liner installed.

BEFORE you install the walls, place an A-frame ladder inside the pool perimeter so you can get back out when the wall is fully assembled. Uncoil the pool wall and walk it around the interior of the

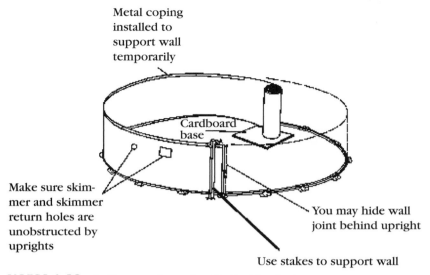

Metal coping installed to support wall temporarily

Cardboard base

Make sure skimmer and skimmer return holes are unobstructed by uprights

You may hide wall joint behind upright

Use stakes to support wall

FIGURE 1-28 **Rolling out the pool wall.** *Sharkline Pools.*

perimeter, setting it into the bottom rail (Fig. 1-29). As you insert the wall, be sure the skimmer and discharge openings are not aligned with any of the base plates. If they are, it means that when you install the uprights, these openings will be covered by the uprights. If you do not have several helpers, you may need some tall landscaping stakes to temporarily brace the wall. As you work, apply the slotted tubes to the top of the pool wall (and the special retainer rail if you are using a beaded liner) as shown in Fig. 1-30. This will protect your hands from the sharp metal and provide additional rigidity to the walls as you assemble them.

When you have completed the circle (or are ready to join wall sections of larger pools), bolt the ends together as shown in Fig. 1-31. Loose fit all bolts and nuts first, then tighten them all. To help you align the holes, insert a screwdriver through the first pair of holes. Remember to put the round head of the bolt on the inside of the pool and the nut on the outside to avoid abrasive surfaces that could cause the liner to fail later. When you have secured the ends firmly together, apply at least three layers of duct tape over the heads of the bolts.

Now install the uprights to the base plates as shown in Fig. 1-32. Oval and rectangular pool supports are installed in much the same way, but they may include additional buttresses and brackets,

FIGURE 1-29 **Installing the pool wall into the base rail.** *Northwest Wholesale.*

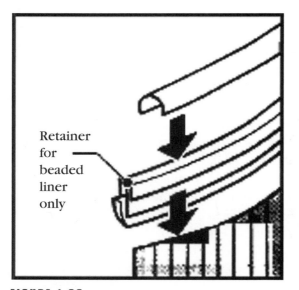

FIGURE 1-30 **Wall top-edge protector and beaded-liner rail installation.** *Cantar/Polyair Corp.*

Retainer for beaded liner only

which are unique to each manufacturer. By learning the steps outlined for a typical round pool, the added component assembly for other shapes will not pose a significant challenge or additional work, but as with other aspects of installation procedures, it is important to follow all manufacturer instructions carefully. Secure the uprights only to the base plates at this point—don't secure them to the top of the wall. If they are too loose to remain upright for now, clip them temporarily to the pool wall with a clothespin or wide-mouth paper clip.

5. **Inspect Your Work** Before hanging the liner, check your work so far. Measure the top of the pool walls across several points to be sure you have a truly round shape. If you're off by a few inches, nudge the base plates in or out with your foot or by gently tapping them with a hammer until all measurements are equal. If large adjustments are needed, you may need to disassemble your work and start over.

Use your carpenter's level again to ensure that the entire structure is level. To be sure the walls are level from one side to the other, hang a string across the diameter of the pool (keep it tight for accurate measurements) and use a line-level. Make any adjustments needed to get the pool level to 1 inch (25 millimeters) all around.

Oval or rectangular pools must be laid out with additional care because the long straight sides are under more

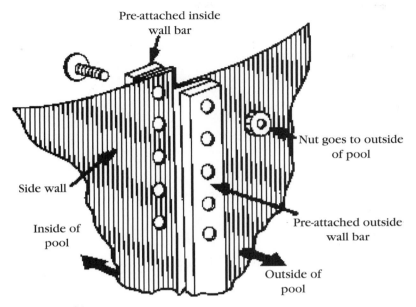

Pre-attached inside
wall bar

Nut goes to outside
of pool

Side wall

Inside of
pool

Pre-attached outside
wall bar

Outside of
pool

FIGURE 1-31 **Bolting the wall ends together.** *Sharkline Pools.*

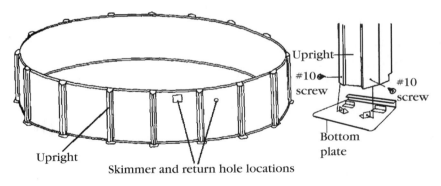

Upright

#10
screw

#10
screw

Bottom
plate

Upright

Skimmer and return hole locations

FIGURE 1-32 **Assembling the uprights.** *Sharkline Pools.*

pressure when the pool is filled than the shorter or rounded sides. Figure 1-33 shows how to measure across typical buttresses to ensure accurate squaring.

If you make any adjustments, examine each base plate to be sure the wall is still firmly inserted and situated on the patio block.

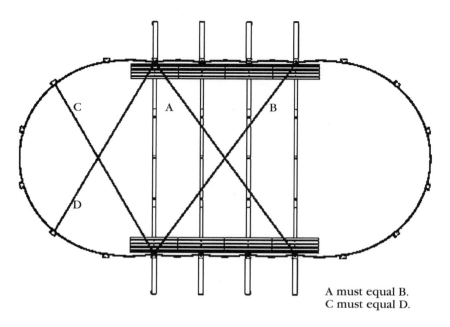

A must equal B.
C must equal D.

FIGURE 1-33 **Squaring the straight sides.** *Sharkline Pools.*

6. **Complete the Bottom Preparation** Before installing the liner, you have to lay the main drain in place if your pool is so equipped. Set the drain unit into the excavation you made earlier. Set a length of PVC flex-pipe into the trench you made, running it under the pool base rail, and then plumbing it to the drain unit. Secure this assembly in place by filling the gaps with dirt and compacting it firmly by hand, then covering it with masonry sand. Make sure the result is level with the surrounding ground. Remove the main drain cover, retaining ring, and gasket, and set these aside until the liner has been installed. Raise the open end of the main drain pipe up and tape it to the top of the pool wall. When you start filling the pool, this will prevent the main drain line from emptying it. Remove the main drain cover, retainer ring, and gasket before installing the liner. If your main drain unit has two gaskets, put a dab of silicone sealant on the lower gasket and lay it over the rim of the main drain, carefully aligning the holes in both. Let this dry before laying the liner over it so it doesn't come loose before you are able to install the retainer ring and the second gasket.

Before hanging the liner itself, you might want to cover the ground with some clean, fine masonry sand to cover any minor pot-holes. Don't use any material that is highly alkaline or acidic because your metal walls will corrode. If your soil is already fairly sandy and smooth, this step won't be necessary. Use a garden hose to lightly dampen the ground to keep it firmly in place while you work and to minimize dust.

To create a smooth, supportive bed for the liner, two additional elements are needed. First, a protective bottom liner should be laid over the entire floor area of the pool. Not all manufacturers insist on this, especially if your soil is smooth, but the added protection is worth the effort and minimal cost. Bottom liners are made of vinyl or thin foam materials, typically provided in rolls that require some cutting and assembly.

Figure 1-34 shows several rows of bottom liner material rolled out and cut to shape. Each section of the material should be joined to the next by seam tape. It may be easier to assemble the bottom liner on your garage floor or your driveway or on another hard surface. Do not overlap the sections, but lay them end to end and join them only with tape. Some bottom liner products recommend a 2-inch (51-millimeter) overlap, so follow the product instructions. Trim the final product to the dimensions of the pool and lay it inside the pool walls. Gladon, the maker of foam bottom liner material, recommends laying out the assembled product before

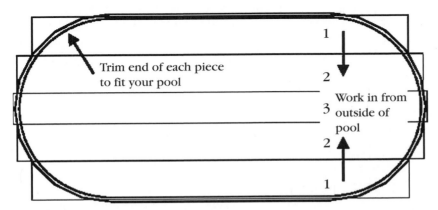

FIGURE 1-34 **Bottom liner installation.** *Gladon Company, Inc.*

installing the pool walls on the bottom rail. Some pool manufacturers specifically suggest adding the bottom liner after the walls are in place, so use your best judgment, but don't do anything that might leave the walls uneven and void the warranty.

Next, the transition area from the wall to the floor of the pool needs to be sloped, not left at a 90-degree angle. Many installers use sand or other neutral fill material to build a *cove* (Fig. 1-35), but it is much easier to use prefabricated cove material (Fig. 1-36). Prefab cove is superior to a handmade version for several other reasons. It won't erode or shift, it is uniform throughout, and it is self-adhesive, so you ensure a solid placement to the pool wall and rail. Install sections of cove around the pool as shown, cutting the last one (use a fine-tooth hacksaw to make a smooth, square cut) slightly larger than the actual space remaining. When you insert this last section, the extra bulk will ensure a tight fit with no gaps.

7. **Hang the Liner** The pool liner itself can now be installed. Take it out of the carton (don't open the box with a knife!), and lay it out in the sun on a clean surface. The warmth of the sun will relax the creases and make the vinyl more supple. While the liner is relax-

FIGURE 1-35 Handmade cove. *Northwest Wholesale.*

ing, punch out the openings in the metal wall for the skimmer and return lines with a hammer and screwdriver (if they are not already cleared), and apply the self-adhesive skimmer gasket to both sides of the opening, "sandwiching" the pool wall. Be sure the bolt holes in the gasket line up with the holes around the skimmer opening.

Carefully deploy the liner around the inside of the pool (Fig. 1-37A and B), hanging the edges over the side of the pool walls (or inserting the bead into the J hook in the retainer rail) and holding it temporarily with clothespins or wide-mouth paper clips. Don't trim off the excess because you may need it later. Insert the liner between the uprights and the pool wall. As you work, inspect all of the vinyl and seams for holes or imperfections that might cause leaks. The round seam should look like a line drawn evenly around the cove. The cross seams should be parallel and not running over the skimmer or return outlet

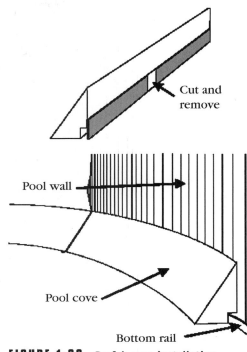

FIGURE 1-36 Prefab cove installation. *Gladon Company, Inc.*

cutouts in the pool walls. The material should lay smooth and snug against the pool wall, cove, and bottom of the pool (Fig. 1-38). Remove any large wrinkles by loosening the fabric at the top, rather than forcing or stretching the material at the bottom. Use a long-handled broom to reach over the pool wall (Fig. 1-40), and smooth out the bottom of the liner against the cove.

Secure your work by replacing the clothespins or wide-mouth paper clips with the inner combing (the slotted rubber gasket that holds the liner on the top of the wall and is then pinched tightly by the coping) and plastic coping along the top edge of the pool wall, starting at the wall bolt area and working around in a circle. When you're done, test yourself against the pros. According to California industry standards, professionally installed liners should remain securely fastened around the top of the pool for at least 1 year and be essentially wrinkle free. Maximum allowable

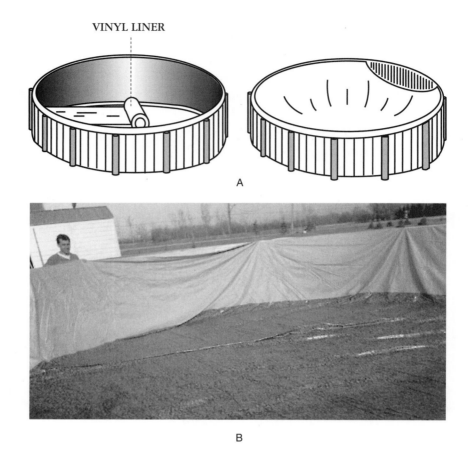

VINYL LINER

A

B

FIGURE 1-37 (A) Rolling out the liner. (B) Hanging the liner. *B: Northwest Wholesale.*

wrinkles should not be deeper than $1/4$ inch (6 millimeters) for folds extending in length 36 inches (1 meter) or less.

Before starting to fill the pool with water, complete the main drain installation (Fig. 1-39). Find the drain under the liner, and gently depress the vinyl into the drain cavity, but not more than 1 inch (25 millimeters) deep. Probe the rim of the main drain unit to find the holes that accommodate the screws that will secure the retainer ring. Poke a hole through the vinyl liner when you find one of these holes. Now place the gasket and retainer ring over the main drain, and screw them in place with one of the stainless steel screws provided. Don't tighten this first screw just yet. It should now be fairly easy to locate each of the other holes and to complete

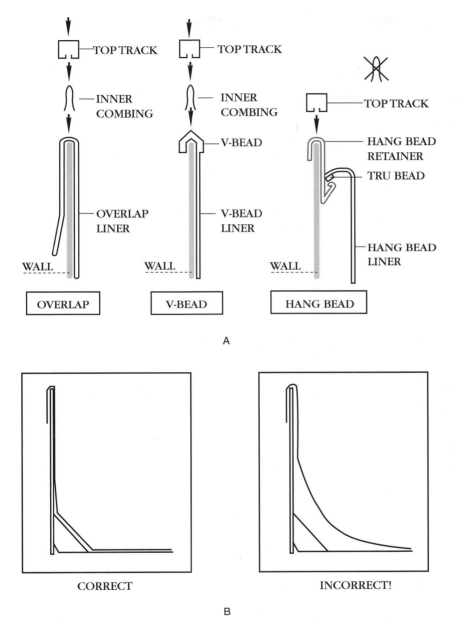

A

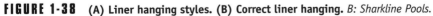

B

FIGURE 1-38 (A) Liner hanging styles. (B) Correct liner hanging. *B: Sharkline Pools.*

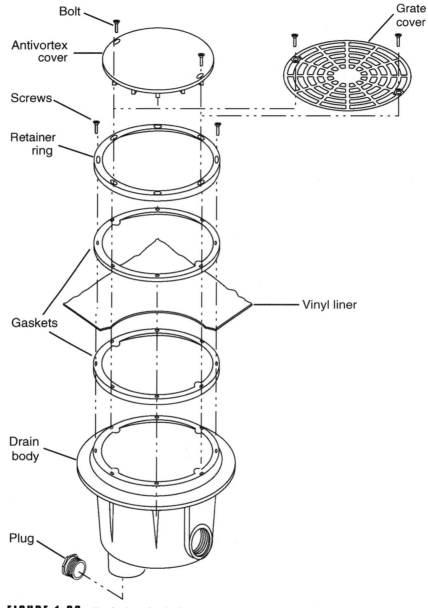

Bolt

Antivortex cover

Grate cover

Screws

Retainer ring

Gaskets

Vinyl liner

Drain body

Plug

FIGURE 1-39 **Typical main drain components.** *Pentair Pool Products, Inc.*

the assembly of the retainer ring and gasket to the main drain unit. When all screws are in place, tighten them all.

Now carefully cut away the liner from inside the circumference of the main drain unit, using a razor knife, always cutting inside the retainer ring. Finish by adding the main drain cover.

8. **Fill the Pool** Start filling the pool with water, and smooth out new wrinkles by removing a section of the coping and adjusting the liner up or down (Fig. 1-40). Work your way around the entire pool. With larger pools, you may have to stop adding water after the first 1 foot (30 centimeters) to give yourself time to finish all adjustments before there's too much weight in the pool and shifting the liner is no longer possible. In any case, don't add more than 6 inches (15 centimeters) of water until the uprights have been fully attached.

To assist with wrinkle removal, you might want to use a vacuum cleaner. Cover the skimmer and return line openings on the outside of the pool wall with cardboard and duct tape; then slide the vacuum hose between the wall and liner and tape it in place. Turn on the vacuum and watch the liner "shrink" into place (Fig. 1-41).

FIGURE 1-40 Smoothing out the liner by hand. *Northwest Wholesale.*

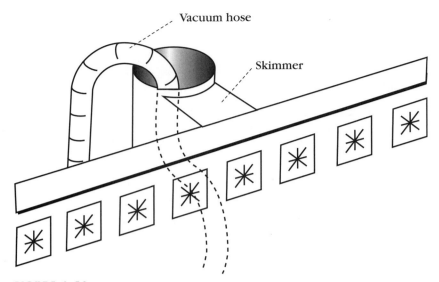

FIGURE 1-41 Removing wrinkles from the liner with a vacuum.

Now secure the uprights to the top of the pool wall, adding the top plates (which hold the actual top rail/cover) as shown in Fig. 1-42. Then secure the top rail all around the perimeter. Don't tighten the screws that secure the top rail to the top plates until all have been installed and you can see that everything fits symmetrically. Now tighten all screws. As you work around the pool, you may need to push the top of each upright in toward the center of the pool to bring it into alignment. This will also help the bottom kick slightly outward, perfecting the circle of the pool.

Some pools have additional resin plastic decorative covers for the uprights or top rail. These typically snap into place. Add them now and continue filling the pool, but not higher than the discharge line opening in the pool wall. The pool may bulge or shift or make some popping noises while filling, but this is normal.

By the way, do not trim off the excess liner. It may be needed for making adjustments later (Fig. 1-43). If you have a beaded liner, this precaution is unnecessary.

9. **Install the Equipment, Plumbing, and Electrical Connections** The plumbing includes the skimmer, main drain, pipes to the equipment area and back to the pool, and return outlet(s). This section

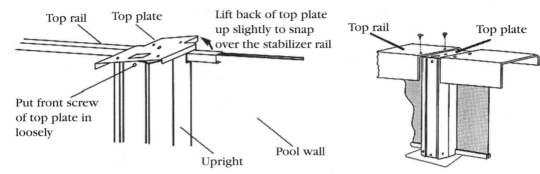

FIGURE 1-42 **Installing the top plate.** *Sharkline Pools.*

FIGURE 1-43 **Liner overlap and typical upright covers.**

will use terms and skills described in the following chapters on plumbing, typical pool equipment, and skidpacks.

If your pool is equipped with a main drain, you have already plumbed it in place. Next, install the skimmer (Fig. 1-44) but not until the water level is just below the location on the pool wall where the discharge port will be installed. If you install the skimmer or discharge port too soon, the liner may not be fully seated

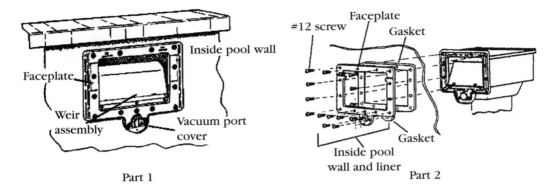

Part 1

Part 2

FIGURE 1-44 Skimmer installation. *Hoffinger Industries/Doughboy Pools.*

into its final position. Be sure all wrinkles are removed and the liner is firmly, fully in place.

Set up your skimmer unit with the weir, basket, and any other components. From the outside of the pool, locate the hole in the wall provided for the skimmer and the small holes provided for the bolts. Punch through the two top bolt holes with a nail or awl. From inside the pool, insert two bolts through (in this order) the faceplate, one of the two skimmer gaskets, the liner, and the pool wall. Attach the second skimmer gasket to the outside of the pool wall by hanging it onto these two bolts, then hanging the skimmer itself on the bolts. Secure the skimmer by adding nuts to the screws, but do not fully tighten them. Punch through the remaining bolt holes, taking care to align the gaskets without cutting them, and finish bolting the skimmer to the pool wall. When all bolts and nuts are in place, tighten them securely.

Use a razor knife to cut the liner material out of the skimmer mouth, taking care to cut only inside the faceplate circumference. Also cut the liner material from the interior of the vacuum port if your skimmer has one.

Next, install the return line discharge port. Find the hole in the pool wall and carefully cut the liner to the same size as that opening. Place a gasket on the return line fitting, and slide it through the wall and liner, from inside the pool. Add another gasket to the fitting from the outside of the pool, and secure the entire assembly in place with the threaded retainer ring that screws onto the round return line fitting. Don't overtighten because these plastic parts can easily crack, but be sure the gaskets are firmly secured. Insert the directional eyeball or other accessories from the inside of the pool.

Plumb a length of flex PVC pipe from the return line fitting to the equipment area. If you have also plumbed a main drain, connect the main drain pipe and the skimmer pipe to a three-port valve in a convenient location at the base of the pool wall. Run another length of pipe from the three-port valve to the pump in your equipment package.

The circulation and filtration equipment is typically provided on a platform as a single unit called a *skidpack*. Figure 1-45 shows a simple unit, set near the pool and plumbed to the pool with a main drain and three-port valve. The water enters the skidpack through the pump and exits from the filter, so plumb the return pipe from the filter to the return line fitting (discharge port) at the pool. Now you are ready to finish filling the pool.

FIGURE 1-45   Skidpack plumbed to pool.

10. **Start the Circulation Equipment** The pump's motor typically is designed with a standard cord and plug to obtain electricity from a normal household wall outlet. Be sure the outlet is equipped with a ground fault circuit interrupter (see Chap. 7) and is weatherproof against exposure to the elements. After you plug in the equipment when the water has flooded the skimmer, you are ready to start the circulation. Some skidpacks are equipped with a timer and on/off switch.

    Check that there is enough water to keep the skimmer flooded, and be sure the weir inside the skimmer is floating unrestricted. Purge air from the system via the filter's air relief valve (see Chap. 5 for more details), and look for steady flow of water at the return line fitting. Add the first dose of sanitizer in accordance with package directions (see Chap. 8), but do it only when the pool is circulating freely to ensure complete distribution of the sanitizer.

    If you have other equipment or want a more permanent installation of your circulation system, hire an electrician to ground any metal within 5 feet (1.5 meters) of the water's edge. The electrician will add light fixtures, which must usually be at least 24 inches (60 centimeters) below the water surface, unless they are sealed, over-the-edge units (see Chap. 7).

11. **Inspect for Leaks** Look around the pool, plumbing, and equipment for leaks. Don't be fooled by moisture left over from the installation or water that may have splashed out of the pool when the air was purged from the equipment upon startup. The gaskets at the skimmer, main drain, and return line fitting will need to get wet and expand, so a slight leak may be normal at first, but it should seal within a few minutes. Most leaks are slow and take some time to be noticed, so take your time with this inspection. Tighten or reseat any components that are loose enough to cause leaks. Repeat the inspection daily for the first week of pool operation because all parts will be settling in place, especially as the pool is put into use.

12. **Add a Ladder, Slide, or Deck** There are so many varieties of these accessories that the best way to describe their installation is to recommend that you follow the manufacturer's instructions. Chapter 7 describes many options. It is best to add these options several

days after you have installed your pool, so that you have had plenty of time to inspect for leaks or defects. Landscaping is also best added after you have used your pool for a few weeks and know more about your use patterns and any problems you may experience with the installation.

13. **Enjoy the Pool** This step seems obvious, but a few more pointers are in order. As noted elsewhere, never allow diving or jumping into an above-ground pool. Most top rails are designed for bearing your weight as you use them to get in or out and if you sit on them, but they are not designed for you to stand on. Finally, don't use the pool if the bottom is not clearly visible. It's easy to bump the bottom when swimming, which can cause an injury, and a cloudy pool is a sign that serious chemical or filtration problems exist.

FAQS: ABOVE-GROUND POOLS

How long will my above-ground or onground pool last?

- With proper care and maintenance, frameless onground pools and soft-sided/framed pools will last at least 10 years; rigid-sided above-ground pools will last twice that long, but the liner will need replacement approximately every 5 years.

Can I put an above-ground or onground pool anywhere?

- The key to placement of all onground and above-ground pools is that the surface be firm and level to within 1 inch (2.5 centimeters). It must also be a surface designed to hold the weight of the water, so building the pool on top of a deck is not advisable. Many manufacturers recommend against setting up the pool on asphalt or concrete because the petroleum in asphalt will eat through the vinyl liner and concrete will cause friction and tears.

Can I move my above-ground pool to a new home?

- Onground pools are designed to be dismantled and moved in an hour or two, but rigid-sided above-ground models will take significantly more time. Draining and removing the vinyl liner of an above-ground pool will probably cause stresses and shrinkage, which would require that it be replaced.

Can an above-ground pool be used the same way an inground pool is used?

- NO! All varieties of above-ground pools, even those with deep ends, are too shallow for diving or jumping. Otherwise, they are put to the same recreational uses.

CHAPTER 2

Basic Above-Ground Pool Plumbing

To understand a pool, you must follow the path of the water. This and all following chapters have been arranged in this manner and you will find it is easy to troubleshoot a pool problem by following this path.

As can be seen in Fig. 2-1, water from the pool enters the equipment system through a main drain on the floor, through a surface skimmer, or through a combination of both main drain and skimmer. It travels to a three-port valve and into the pump, which is driven by the attached motor. From the pump, the water travels through a filter, up to solar panels (if part of the installation), then to the heater, and back to the pool return line.

Skimmers

Small above-ground pools have no skimmer, mid-size pools have one skimmer, and larger pools have more than one skimmer. The purpose of the skimmer, as the name implies, is to pull water into the system from the surface with a skimming action, pulling in leaves, oil, and dirt before they can sink to the bottom of the pool. It also provides a

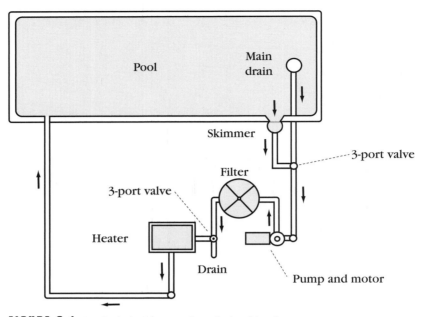

FIGURE 2-1 Typical above-ground pool plumbing layout.

conveniently located suction line for vacuuming the pool. Most skimmers today are molded, one-piece plastic units that mount externally (Fig. 2-2A).

The skimmer is accessed through a cover on top or though the faceplate (Fig. 2-2B). Most skimmers for above-ground pools are separate units that hang on the edge of the pool (in the water or outside of it). Some pools use a flat, vertical skimmer that has no basket but skims the surface and pulls any floating debris to a plastic screen. Some skimmers have a cartridge filter built in.

The water pours over a floating weir (Fig. 2-2C) that allows debris to enter, but when the pump is shut off and the suction stops, the weir floats into a vertical position, preventing debris from floating back into the pool. Some skimmers have no such weir (although spring-loaded weirs are available that can be fitted into any skimmer mouth) and use a floating barrel (Fig. 2-2D) as part of the skimmer basket. The purpose of the basket is to collect leaves and large debris so they can then be easily removed.

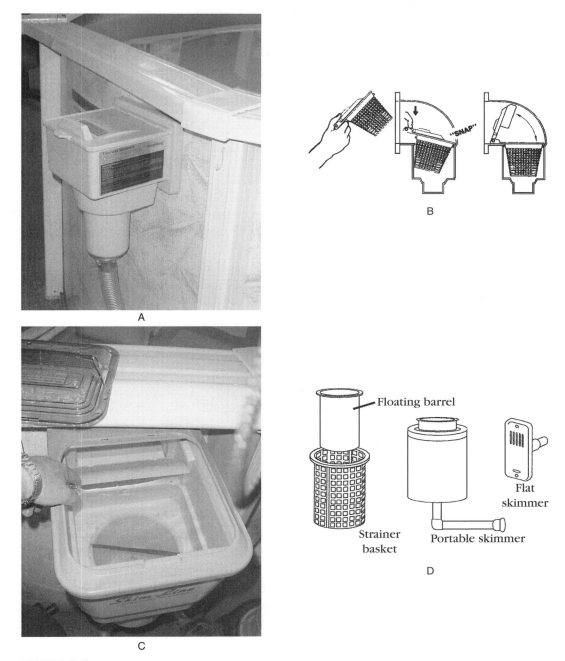

FIGURE 2-2 (A) Typical externally mounted skimmer. (B) Front-loading skimmer. (C) The skimmer weir. (D) Floating and flat skimmers. *B: Jacuzzi Bros., Little Rock, Ark.*

The disadvantage of both types of weirs is that leaves can cause them to jam in a fixed position, thus preventing water from flowing into the skimmer. When this happens, the pump will lose prime (water flow) and run dry, causing damage to its components. Therefore, during windy periods it might be better to remove the weir from the skimmer to prevent such problems.

By the way, you should exercise care when working around the skimmer when the pump is on. I have nearly had fingers broken when placing my hand over a skimmer suction opening and have lost various pieces of equipment, T-shirts, bolts, and plastic parts, which invariably end up clogged in the pipe at some turning point where leaves, hair, and debris later catch and close off the pipe completely. Keep small objects away from the skimmer opening when the basket is removed and especially keep your hands from covering that suction hole.

Some skimmers have odd-sized ports or tight angles that can't accommodate your vacuum hose. In these cases, a special cover plate (Fig. 2-3) can create a generic adapter.

Main Drains

The main drain on an above-ground pool has one port that feeds a pipe to the pump. A three-port valve at the pump allows you to balance the suction between the skimmer and optional main drain. When cleaning the pool with a vacuum or suction-side automatic cleaner, you may want to set the valve in the position that directs all suction to the skimmer. If the bottom of the pool is especially dirty, you can set the valve to direct all of the suction to the main drain and brush the dirt toward it. More about that technique, and other ways to use the balance between the main drain and skimmer, is found in Chap. 9, "Cleaning and Servicing."

If you are offered the option of adding a main drain to your pool, I highly recommend it. Most above-ground pools have only one suction port (via the skimmer) and one discharge port for returning water to the pool, often located within a few feet of each other. This close proximity of the suction and discharge ports means that the end of the pool farthest from skimmer will inevitably receive less circulation and therefore be more likely to develop algae and other problems. Adding

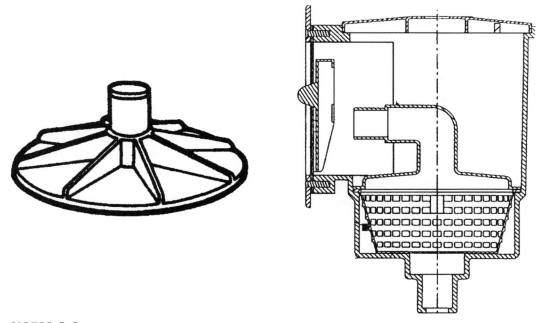

FIGURE 2-3 Skimmer vacuum adapter plates.

a main drain (or a second discharge port farther from the skimmer) helps to improve circulation significantly.

Because above-ground pools are relatively shallow, strong suction can create a whirlpool effect. To prevent this, many main drains are fitted with antivortex drain covers that are slightly dome-shaped with the openings located around the sides of the dome (Fig. 2-4). Main drain covers for deeper pools are flat with the openings on top.

In any case, the drain area is covered by a grate, usually 8 to 12 inches (20 to 30 centimeters) in diameter, that screws or twist-locks into a ring that has been built into the pool bottom.

General Plumbing Guidelines

Good news: The vast majority of plumbing for above-ground pools is made of easy-to-use PVC plastic. Some larger pools with older equip-

FIGURE 2-4 Main drain antivortex cover and standard cover.

ment and natural gas supply lines to pool heaters may involve galvanized plumbing. Copper pipes may be found in heaters or other ancillary equipment, so we will examine the best ways to work with all three types of plumbing material.

Before proceeding to specific instructions on working with PVC plastic, galvanized, or copper plumbing, here are a few general guidelines that I think are important regardless of the material you are using.

Measure the pipe run carefully, particularly if you are repairing a section between plumbing that is already in place. In measuring, remember to include the amount of pipe that fits inside the connection fitting, usually about $1\frac{1}{2}$ inches (3.8 centimeters) at each joint (Fig. 2-5).

When working on in-place plumbing, support your work by building up wood or bricks under the pipe on each side of your work area. This prevents vibration as you cut, which can damage pipes or joints further down the line. Also, unsupported pipe sags and binds when you cut it. That is, as you cut, it pinches the saw blade making cutting difficult, and straight, clean cuts impossible!

Threaded fittings are obvious and simple; however, leaks occur most often in these connections. The key is to carefully cover the male

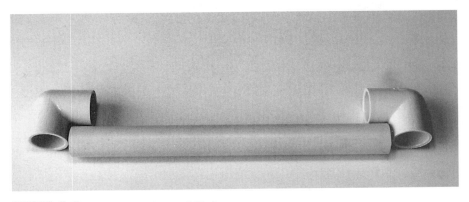

FIGURE 2-5 Pipe measuring and fitting.

threads with Teflon tape and to tighten the fitting as far as possible without cracking.

Teflon tape fills the gaps between the threads to prevent leaking. Apply the tape over each thread twice, pulling it tight as you go so you can see the threads. Apply the tape clockwise (Fig. 2-6) as you face the open end of the male threaded fitting. If you apply the tape backwards, when you screw on the female fitting, the tape will skid off the joint. Try it both ways to see what I mean and you will only make that mistake once.

Another method of sealing threads is to apply joint stick or pipe dope. These are odd names for useful products that are applied in similar ways (Fig. 2-7). Joint stick is a crayon-type stick of a gum-like substance that works like Teflon tape. Rub the joint stick over the threads so that the gum fills the threads. Apply pipe dope the same way. The only difference is that dope comes in a can with a brush and is

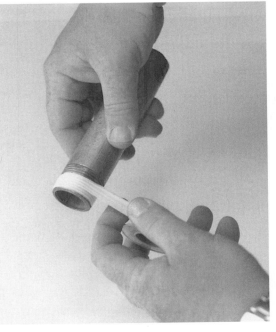

FIGURE 2-6 Correct application of Teflon tape.

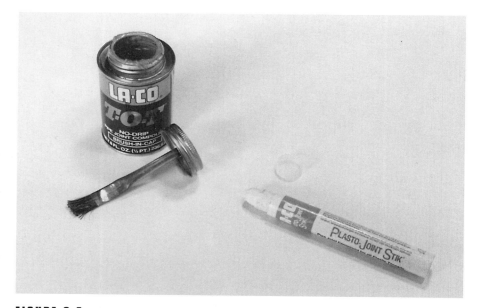

FIGURE 2-7 Pipe dope and joint stick.

slightly more fluid than joint stick. The key to success with joint stick or pipe dope is to apply it liberally and around all sides of the male threaded fitting, so that you have even coverage when you finally screw the fittings together. Some product will ooze out as you tighten the fittings together, but that proves that you have applied enough.

I use Teflon tape because I know upon application that it is an even and complete coverage of the threads. Pipe dope or joint stick might not apply evenly or can bunch up when threading the joint together. If you use dope or stick, be sure it is a nonpetroleum-based material (such as silicone). Petroleum-based products will dissolve plastic over time, creating leaks.

When working with PVC pipe and fittings, tighten threaded fittings with channel lock pliers of adequate size to grip the pipe. Using pipe wrenches usually results in application of too much force and cracking of the fittings. Save the pipe wrenches for copper or galvanized plumbing. If you don't have pliers large enough for the work and must use a pipe wrench, tighten the work slowly and gently—not bad advice with copper either, because copper is soft and will bend or crimp if too much force is applied.

To avoid slipping off the work and damaging fittings, always tighten with channel locks or wrenches into the jaw, not away from it (Fig. 2-8). Another way to think of it is into their base, not their head.

Now a word about pipe cutters (Fig. 2-9). Made for PVC or metal pipe, these are adjustable wrench-like devices that have cutting wheels. You lock the device around the pipe and rotate it, constantly tightening it as you go, until the pipe is cut. They provide the straightest, cleanest cut of all. However, in pool work you will be dealing mostly with 1½- to 2-inch-diameter (40- to 50-millimeter) pipe in close quarters. You rarely have the luxury of enough space to get around the entire pipe, and these cutters take far longer than a good, fresh hacksaw

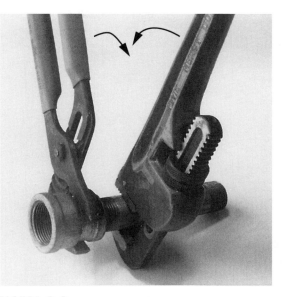

FIGURE 2-8 Correct wrench use.

blade. Use them if you like, but I think you will soon abandon them in favor of simple old Mr. Hacksaw.

The most important advice I can give you on odd-job or tight-quarters plumbing, indeed on any of this plumbing, is to ask questions. There are so many different, unique fittings and fixtures for cutting, joining, and repairing plumbing that they can fill a book of their own, and it would still be out of date because of constant revision and new products. So if you run across a tough connection of odd pipe or different materials, ask questions at your local plumbing supply house. Most of the counter help is knowledgeable and willing to advise you because their advice sells their products. A good idea is to take snapshots of the job (or bring the materials in with you if you must cut them out anyway) so they will thoroughly understand your specific needs.

PVC and Flexible PVC Plumbing

If you played with Tinker Toys, Lego blocks, or Lincoln Logs as a kid, you will find working with PVC plumbing literally, well, a snap.

Pool plumbing is prepared with plastic or metal lengths of pipe and connection fittings that join those lengths together. The pipe acts as

FIGURE 2-9 Pipe cutters.

the *male* which fits and is glued into the *female* openings of these connection fittings. Alternatively, connection is made by each side having threads, joined by screwing them together. The plastic pipe used is PVC (polyvinyl chloride) and it is manufactured in a variety of different strengths depending on the intended use.

To help identify the relative strength of PVC, it is labeled by a *schedule* number; the higher the number, the heavier and stronger the pipe. Pool plumbing is done with PVC schedule 40. Some gas lines are plumbed with PVC schedule 80.

PVC is designed to carry unheated water (under 100°F). CPVC is formulated to withstand higher temperatures for connection close to (or in some cases directly to) a pool heater.

Ultraviolet (UV) light from the sun causes PVC to become brittle over time, losing its strength under pressure and creating cracks.

Chemical inhibitors are added to some PVC to prevent this, the most common and cheapest being simple carbon black (which is why plastic pumps and other pool equipment is often black). Another common preventive measure is to simply paint any pipe exposed to sunlight.

PVC pipe is manufactured in a flexible version, making plumbing easier in tight spaces. Flex PVC is available in colors for cosmetic purposes and has the same characteristics and specifications as rigid PVC of the same schedule and size.

Many of the equipment packages designed for above-ground pools are plumbed with flex pipe and preplumbed threaded connectors. Be careful when setting up this type of plumbing because tight bends can easily kink (Fig. 2-10) and restrict water flow. Figure 2-11 shows long, curving lengths of flex PVC, which may seem excessive but actually help to prevent kinks.

All pipe is measured by its diameter, expressed in inches or millimeters. Typically pool plumbing is done with $1\frac{1}{2}$- or 2-inch (40- or 50-millimeter) pipe, referring to the interior diameter (the diameter of the pipe that is in actual contact with the water). The exterior diameter of the pipe differs depending on the material. For example, the exterior diameter of 2-inch PVC pipe is greater than that of 2-inch copper pipe because the PVC pipe walls are thicker than those of copper.

FIGURE 2-10   Kinks in flex PVC plumbing.

FIGURE 2-11   Flex PVC plumbing runs.

All pipe is connected with *fittings* (Fig. 2-12). Fittings allow connection of pipe along a straight run (called *couplings*), right angles (called *90-degree* couplings or *elbows*), 45-degree angles, T fittings,

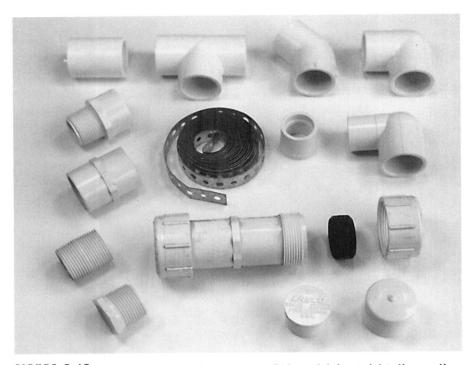

FIGURE 2-12 Pipe connection fittings. Top row (left to right): straight slip coupling, T coupling, 45-degree slip coupling, 90-degree slip coupling. Second row: mip, fip, plumber's strap tape, reducer bushing, 90-degree slip street coupling. Third row: close nipple, compression coupling. Bottom row: male threaded plug, male slip plug, female slip cap.

and a variety of other formats. In the case of PVC, such fittings are most often smooth-fitted and glued together (called *slip* fittings).

Some fittings are threaded (called *threaded* fittings) with a standard plumbing thread size so they can be screwed into comparable connecting fittings in pumps or other plumbing parts. National Pipe Thread (NPT) standards are used in the United States so different products of various materials by different manufacturers will all work together. The NPT standard includes a slight tapering between the male and female connections. The importance of this is that because of this taper, it is easy to overtighten plastic threaded fittings and crack them. Great Britain, Europe, and Asia not only operate on metric measurements, but also have their own unique thread standards. Fittings with male (external) threads are called *mip* and fittings with female (internal) threads are called *fip*. If one side of the fitting is mip and one side is slip, you order it as *mip by slip*, and so on.

The equipment for above-ground pools is generally located adjacent to the pool itself, so extended lengths of pipe are not necessary, but where they exist, the long runs of pipe will be underground. Sometimes, however, horizontal runs will be under a house or deck or over a slope where support is needed. In this case, pipe should be supported every 6 to 8 feet (2 to 2.5 meters), hung with plumber's tape to joists or supported with wooden bracing. PVC does not require support on vertical runs because of its stiffness, but common sense and local building codes might require strapping it to walls or vertical beams to keep it from shifting or falling over. Remember, the pipe becomes considerably heavier when it is filled with water and might vibrate along with pump vibration.

Plumbing Methods

RATING: EASY

The concept of joining PVC pipe involves welding the material together by using glue that actually melts the plastic parts to each other. In truth, each joint will have an area that is slightly tighter than the rest. In the tightest parts, this welding actually occurs. In the remainder, the glue bonds to each surface and itself becomes the bonding agent. Obviously the strongest part of each joint is the welded portion; but in either case, the key is to use enough glue to ensure total coverage of the surfaces to be joined.

Following is the correct procedure for plumbing with PVC (Fig. 2-13):

1. **Cut and Fit** Cut and dry fit all joints and plumbing planned. It is easy to make mistakes in measuring or cutting and sometimes fittings are not uniform so they don't fit well. Dry fitting ensures the job is right before gluing. If you need the fitting and pipe to line up exactly for alignment with other parts, make a line on the fitting and pipe (Fig. 2-13A) with a marker when dry fitting so you have a reference when you glue them together.

2. **Sand** Lightly sand the pipe (Fig. 2-13B) and inside the fittings so they are free of burrs. The slightly rough surface will also help the glue adhere better.

3. **Prime** You might need to apply a preparation material, called *primer*, to the areas to be joined before gluing (Fig. 2-13C). Some PVC glues are solvent/glue combinations and no primer is required. In some states, however, use of primer might be required by building code, so

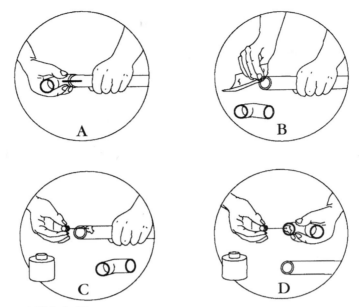

FIGURE 2-13 Step-by-step PVC plumbing.

check that before selecting an all-in-one product. If you are using primer, apply it with the swab provided to both the pipe and the inside of the fitting. Read and follow the directions on the can.

TOOLS OF THE TRADE: PVC PLUMBING

The supplies and tools you need for PVC plumbing are

- Hacksaw with spare blades (coarse: 12 to 18 teeth per inch or 2.5 cm)
- PVC glue and primer
- Cleanup rags
- Fine sandpaper
- Teflon tape or joint stick
- Waterproof marker

---

4. **Glue** Before gluing, be ready to fit the components together quickly because PVC glue sets up in 5 to 10 seconds. Apply glue to the pipe and inside of the fitting (Fig. 2-13D).

5. **Join** Fit the pipe and fitting together, duplicating your dry fit, and twist about a half turn to help distribute the glue evenly, realigning the lines drawn on the pipe and on the fitting. If using flexible PVC, because it is made by coiling a thin piece of material and bonding it together, do not twist it clockwise. This can make the material swell and push the pipe out of the fitting. Get in the

habit of twisting all pipe counterclockwise (even though it makes no difference with rigid PVC) and you will never make that mistake.

6. **Seal** With rigid PVC, hold the joint together about a minute to ensure a tight fit; about two minutes with flex PVC. Although the joint will hold the required working pressure in a few minutes (and long before the glue is totally dry), allow overnight drying before running water through the pipe to be sure. I have seen demonstrations with some products (notably Pool-Tite solvent/glue) where the gluing was done underwater, put immediately under pressure, and it held just fine. I don't, however, recommend this procedure as I have gone back on too many plumbing jobs to fix leaks a few weeks later because I hastily fired up the system after allowing only a few minutes drying time.

FAQS: BASIC PLUMBING

Do I need a main drain for an above-ground pool?

• You don't need a main drain, but the more circulation that exists throughout the pool, the less likely you will be to have problems with cloudy water or algae. If you don't have a main drain, consider adding a second return outlet (or a second skimmer on large pools).

Can I bury the plumbing and locate the pool equipment remotely?

• Plumbing for above-ground pools can be buried as is any other pool plumbing. However, be sure to follow local building codes and bury pipes deep enough to avoid freeze damage. If your equipment will be more than 30 feet (9 meters) from the pool, consult a pool professional to be sure the pump is adequate.

Do I need to drain my pool every time I work on the plumbing?

• Your pool should be fitted with shutoff valves at the suction and discharge sides of the pool to isolate the plumbing and equipment when you are working on it, so draining won't be necessary.

Is PVC pipe the same as flex PVC pipe?

• Yes. Working with either rigid or flexible PVC is identical and requires the same tools, skills, and precautions.

TRICKS OF THE TRADE: PVC PLUMBING

1. Make all threaded connections first, so if you crack one while tightening it can be easily removed. Then glue the remaining joints to the threaded work.

2. When cutting PVC pipe, hacksaw blades of 12 teeth per inch (2.5 centimeters) are best, particularly if the pipe is wet (as when making an on-site repair). Finer blades will clog with soggy, plastic particles and stop cutting. Use blades of 10 inches (25 centimeters) in length. They wobble less than 12- or 18-inch blades during cutting. In all cases, the key is a fresh, sharp blade. For the few pennies involved, change blades in your saw frequently rather than hacking away with dull blades—you'll notice the difference immediately.

3. No matter how careful you are, you will drip some glue on the area or yourself. That's why I always carry a supply of dry, clean rags to keep myself, the work area, and the customer's equipment clean of glue.

4. Try to make as many *free* joints as possible first. By that I mean the joints that do not require an exact angle or which are not attached to equipment or existing plumbing. The free joints are those that you can easily redo if you make a mistake. Do the hard ones last—those that commit your work to the equipment or existing plumbing and cannot be undone without cutting out the entire thing and starting over.

5. Use as much glue as you need to be sure there is enough in the joint. It's easier to wipe off excess glue than to discover that a small portion of the joint has no glue and leaks.

6. Practice. PVC pipe and fittings are relatively cheap, so make several practice joints and test them for leaks in the shop before working on someone's equipment in tight quarters in the field.

7. Flexible PVC is the same as rigid, but when you insert the pipe into a fitting, hold it in place for a minute or longer because flex PVC has a habit of backing out somewhat, causing leaks.

8. In cold weather, more time is required to obtain a pressure-tight joint, so be patient and hold each joint together longer before going on to the next.

9. Bring extra fittings and pipe to each job site. Bring extras of the types you expect to use, as well as types you don't expect to use, because you just might need them. Nothing is worse than completing a difficult plumbing job and being short just one fitting, or needing to cut out some of your work and not having the fittings or a few feet of pipe to replace them. It is often several miles back to the office or the nearest hardware store to grab that extra fitting that should have been in your truck in the first place. Bring extra glue, sandpaper, and rags, too.

Copper Plumbing

Copper plumbing is quickly disappearing from all pools and spas for a number of reasons. Unlike PVC, metal plumbing such as copper will corrode, especially in the presence of caustic pool and spa chemicals moving at high speed and under pressure through the pipes. Copper plumbing is also more difficult to install and repair, and it has become extremely expensive in recent years.

Copper was more recently still in use where dissipation of heat was important, such as in plumbing directly connected to the heater. Stainless steel heat risers, CPVC, and threaded galvanized plumbing have replaced that function, making copper plumbing for pools and spas obsolete.

Copper plumbing is prepared and installed using many of the same techniques as PVC, but the plumbing process is called *sweating*, and it requires a blowtorch, solder, and other special tools. If you find you need these skills for a unique installation, call a pro or refer to *The Ultimate Pool Maintenance Manual* for all of the steps and Tricks of the Trade for working with copper plumbing.

The one type of copper fitting you may encounter in above-ground pool equipment is the *compression fitting* (Fig. 2-14). This fitting is used in small-diameter (1/4- to 1/2-inch or 6- to 13-millimeter) pipe, often inside the heater (see the pressure switch section in Chap. 6). The compression nut is placed on the pipe followed by the compression ring. The pipe is placed inside the opposing fitting and when the nut is screwed onto that fitting, the compression ring tightens around the pipe and seals it.

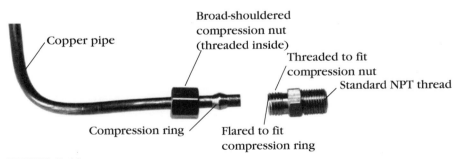

Copper pipe

Broad-shouldered
compression nut
(threaded inside)

Threaded to fit
compression nut

Standard NPT thread

Compression ring

Flared to fit
compression ring

FIGURE 2-14 Copper compression fittings.

CHAPTER 3

Advanced Plumbing Systems

Perhaps the fastest growing segment of the pool industry is in advanced plumbing systems, including automated valves, solar heating, and the use of space-age materials. In this chapter, I discuss some of the more common applications of advanced plumbing and the maintenance of these systems.

Manual Three-Port Valves

The design of three-port valves takes water flow from one direction and divides it into a choice of two other directions. Picture a Y, for example, with the water coming up the stem, then a diverter allows a choice between one of two directions (or a combination thereof). Conversely, the water flow might be coming from the top of the Y, from two different sources, and the diverter decides which source will continue down the stem or mixes some from each together.

Figure 3-1 shows a typical Y or three-port valve, which is very common and easy to use and maintain. Several manufacturers make similar units based on the same concept.

Operation

Whether the three-port valve is Noryl plastic (a type of PVC) or brass, the concept is the same with them all. A housing (Fig. 3-1, item 1),

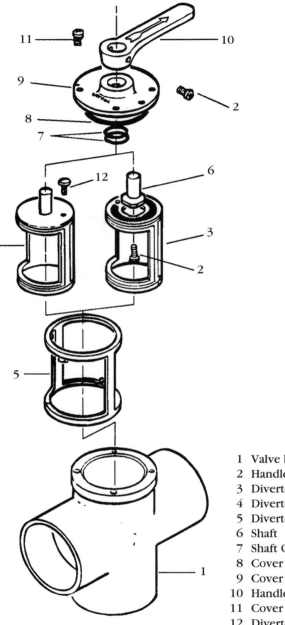

1 Valve body
2 Handle screw
3 Diverter (for separate shaft)
4 Diverter (with shaft built in)
5 Diverter seal gasket
6 Shaft
7 Shaft O-ring
8 Cover O-ring
9 Cover
10 Handle
11 Cover screw
12 Diverter stop screw

FIGURE 3-1   **Construction of a typical three-port valve.** *Pentair Pool Products, Inc.*

built to 1½- or 2-inch (40- or 50-millimeter) plumbing size, houses a diverter (Fig. 3-1, item 3 or 4) that is moved by a handle (Fig. 3-1, item 10) on top of the unit.

Construction

The diverter is surrounded by a custom-made gasket (Fig. 3-1, item 5) so that no water can bypass the intended direction. A valve with this type of diverter and gasket is called a *positive seal* valve.

Some valves, for use where such water bypass is not considered a problem, have no such gaskets and, in fact, the diverter is designed more like a shovel head than a barrel. These divert most of the water in the desired direction with a lesser amount going in the other direction. These are called *nonpositive* valves.

The diverter is held in the valve housing by a cover (Fig. 3-1, item 9) that attaches to the housing with sheet metal screws (Fig. 3-1, item 11), and is sealed with an O-ring (Fig. 3-1, item 8) to make it watertight.

Notice that besides the four screw holes in the cover and housing, there is a fifth hole in the cover that corresponds to a post on the housing. This is to ensure that the cover lines up properly with the housing, because on the underside of the cover are specially molded stops. A small screw (Fig. 3-1, item 12) on top of the diverter hits these stops molded into the underside of the cover. This allows the diverter to be turned only 180 degrees (one-half turn), in either direction, ensuring that the diverter stops turning when facing precisely one side or the other.

This screw is removed when the valve is motorized because the motors only rotate in one direction and are already precise in stopping every half-turn. Small machine screws (Fig. 3-1, item 2) hold the handle on the shaft. A hole in the center of the cover allows a shaft from the diverter to attach to the handle for manual operation of the valve. To make the shaft hole in the cover watertight, two small O-rings (Fig. 3-1, item 7) slide on the shaft in a groove under the cover.

Maintenance and Repair

RATING: EASY

As the simple parts suggest, there's not too much that can go wrong with manual three-port valves.

INSTALLATION

Noryl three-port valves are glued directly to PVC pipe using regular PVC cement like any other PVC fitting. Care should be exercised not to use too much glue, as excess glue can spill onto the diverter and cement it to the housing. Excess glue can also dry hard and sharp, cutting into the gasket each time the diverter is turned, creating leaks from one side to the other.

Brass valves are sweated onto copper pipe like any other fitting. Be sure to remove the diverter when sweating so the heat doesn't melt the gasket.

LUBRICATION

For smooth operation, the gasket must be lubricated with pure silicone lubricant. Vaseline-like in consistency, this lubricant cannot be substituted. Most other lubricants are petroleum-based which will dissolve the gasket material and cause leaks. Lubrication should be done every six months or when operation feels stiff. This preventive maintenance is particularly important with motorized valves because the motor will continue to fight against the sticky valve until either the diverter and shaft break apart or, more often, an expensive valve motor burns out.

Lubrication is the most important maintenance item with any three-port valve. When the valve becomes stiff to turn it places stress on the shaft. On older models, the shaft is a separate piece that bolts onto the diverter (Fig. 3-1, items 2, 3, and 6). Particularly on these models, but actually on any model, the stress of forcing a sticky valve will separate the stem from the diverter. If this happens, turning the handle and stem does not affect the diverter. To repair this, remove the handle and cover, pull out the diverter, and replace it with a one-piece unit (Fig. 3-1, item 4). If the gasket looks worn, replace it and lube it generously before reassembly.

REPAIRS

Few things go wrong with these valves, but the breakdowns that do occur are annoying and recurrent. Before disassembling any valve, check its location in relation to the pool water level. If it is below the water level, opening the valve will flood the area. You must first shut off both the suction and return lines. When installations are made below water level, they are usually equipped with shutoff valves to isolate the equipment for just such repair or maintenance work. If the valves are above the water level, you will need to reprime the system after making repairs (see the section on priming).

Leaks are the most common failure in these valves. The valve will sometimes leak from under the cover. Either the cover gasket is too compressed and needs replacement or the cover is loose. The cover is attached to the Noryl valve housing with sheet metal screws. If tightened too much, the screw strips out the hole and you will be unable to tighten it. The only remedy is to use a slightly larger or longer screw to get a new grip on the plastic material of the housing. Be sure to use stainless steel screws or the screw will rust and break down, causing a new leak. If new screws have already been used and there is not enough material left in the housing for the screw to grip, you must replace the housing. I have managed to fill the enlarged hole with super-type glue or PVC glue and, after it dries, replace the screw. These repairs usually leak and are only temporary measures. You can also fill the hole with fiberglass resin, which usually lasts longer.

Leaks also occur where the shaft comes through the cover. Remove the handle and cover and replace the two small O-rings. Apply some silicone lubricant to the shaft before reassembly. This lubricates the operation of the valve, decreasing friction that can wear out the O-rings. The lubricant also acts as a sealant.

Leaks can occur where the pipes join to the housing ports. In this case, the only solution is replacing the housing. I have tried to reglue the leaking area by removing the diverter and gluing from both inside and outside of the joint. This has never worked! Try if you will, but I think you'll be wasting your time.

Finally, leaks occur inside the valve with no visible external evidence. By this I mean that water is not completely diverted in the intended direction, but slips past the diverter seal to the closed side of the valve. The cause might be a diverter that is not aligned precisely toward the intended port. Remove the diverter and make sure the shaft has not separated or become loose from the diverter.

Unions

An improvement for installing equipment is the plumbing union. When you need to repair or replace a pump, filter, or other equipment that is plumbed into the system, you must cut out the plumbing and replumb upon reinstallation. The concept of the union is that when you remove a particular piece of equipment, you need only unscrew the plumbing and reinstall it later the same, simple way.

Although unions add a few dollars to your initial installation, they allow you to easily remove and replace equipment without doing any new plumbing. Unions, like other plumbing, are made of plastic or metal in standard diameters and are adapted to plumbing like any other component (gluing, threading, or sweating).

Figure 3-2 shows a typical plumbing union. A nut is placed over the end of one pipe, then male and female fittings (called shoulders) are plumbed onto each end of the pipes to be joined. As can be seen, the joint is made by screwing the nut down on the male fitting. Teflon tape or other sealants are not needed as the design of the union prevents leaking (either by the lip design as shown or by use of an O-ring seated between the shoulders).

Unions are made for direct adaptation to pool equipment, where the pipe with the nut and female shoulder is male-threaded at its other end for direct attachment to the pump, filter, or any other female

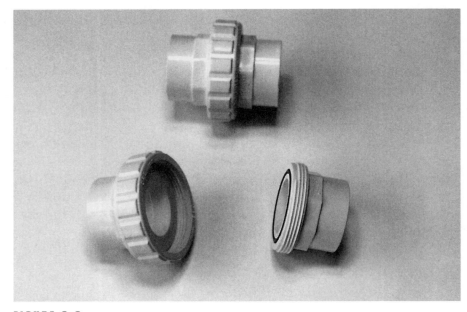

FIGURE 3-2   Plumbing union.

threaded equipment. Then, only the male shoulder need be added to the next pipe and the piece of equipment can be screwed into place.

Gate and Ball Valves

Gate and ball valves are designed to shut off the flow of water in a pipe and are used to isolate equipment or regulate water flow. You might see these in systems where the equipment is installed below the water level of the pool. Without them, when you open or remove a piece of equipment or plumbing you will flood out the neighborhood.

There are basically two types of shutoff valves. Figure 3-3B shows the gate valve. As the name implies, it has a disc-shaped gate inside a housing that screws into place across the diameter of the pipe, shutting off water flow. A variation of this is the slide valve (Figs. 3-3A and 3-4), where a simple guillotine-like plate is pushed into place across the diameter of the pipe.

The other design is the ball valve (Fig. 3-3A), where the valve housing contains a ball with a hole in it of similar diameter as the pipe. A handle on the valve turns the ball so the hole aligns with the pipe, allow-

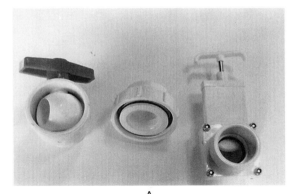

A

B

FIGURE 3-3   Ball (A left), slide (A right), and gate (B) valves.

ing water flow, or aligns across the pipe, blocking flow. In each of these designs, flow can be controlled by degree as well as total on or total off.

The gate valve is operated by a handle that drives a worm screw-style shaft inside a threaded gate. If the gate sticks and too much force is applied to the handle, the screw threads will strip out, making the valve useless. The valve cap (also called the *bonnet*) can be removed

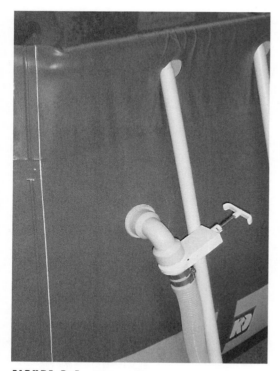

FIGURE 3-4 Slide valve in use.

(unscrewed) and the drive gear and gate can be removed and repaired or replaced without removing the entire valve housing. Most plumbing supply houses sell these replacement guts, but the parts from one manufacturer are not interchangeable with another.

Check Valves

The purpose of the check valve is to check the water flow—to allow it to go only in one direction. The uses are many and will be noted in each equipment chapter where they are employed; however, three common uses are

- In heater plumbing (to keep hot water from flowing back into the filter)

- With chlorinators to keep the flow of caustic chemicals moving in the desired direction

- In front of the pump when it is located above the pool water level (to keep water from flowing back into the pool when the pump is turned off)

There are two types of check valves (Fig. 3-5). One is a *flapper gate* (also called a *swing gate* valve) and the other is a *spring-loaded gate*. The flapper style opens or closes with water flow, while the spring-loaded style can be designed to respond to certain water pressure. Depending on the strength of the spring, it might require one, two, or more pounds of pressure before the spring-loaded gate opens. As with other valves, check valves are made of plastic or metal in standard plumbing sizes and are plumbed in place with standard glue, thread, or sweat methods.

A problem I have encountered with swing gate-type check valves is that the gate comes off the hinge from obstruction damage. In this case, the valve must be replaced. The spring-loaded valve rarely breaks.

The only real weakness of all check valves is that they clog easily with debris, remaining permanently open or permanently closed.

FIGURE 3-5 In-line swing gate, in-line spring-loaded gate, and 90-degree spring-loaded check valves.

Because of the extra parts inside a spring-loaded check valve, they are more prone to failure from any debris allowed in the line. If the valve is threaded or installed with unions, it is easy to remove it, clear the obstruction, and reinstall it.

Another solution is to use the 90-degree check valve. This valve allows you to unscrew the cap, remove the spring and gate, remove any obstruction, and reassemble. Be careful not to overtighten the cap—they crack easily on older models; newer models are made with beefier caps to prevent this problem. These units have an O-ring in the cap to prevent leaks. It is wise to clean these out every few months (or more frequently, depending on how dirty the pool normally gets) and lubricate the gate (using silicone lube only).

These valves are great because of their ease of cleaning and repair, but I don't recommend using them unless you need to make the 90-degree turn in your plumbing anyway. Otherwise you are adding more angles to your plumbing, which restricts water flow unnecessarily.

Some check valves are made of clear PVC, which allows you to see if they are operating normally.

Solar Heating Systems

Solar heating systems are discussed here in the advanced plumbing chapter because it is the plumbing of these systems that presents the most challenges.

Certainly an entire book could be written about solar heating and installations. This is particularly true in today's market where new technologies, alloys, and plastics are being used to manufacture solar panels. Nevertheless, many repairs and some installations are simple, and the plumbing is the same as regular pool plumbing, so leak repair is easy. Here are some guidelines for approaching solar heating systems.

Types of Solar Heating Systems

The function of the solar panel is to absorb heat from the sun which is transferred to the liquid as it passes through. Designs and materials are hotly debated (pun intended) between various manufacturers, but efficiency of a solar heating system is less a factor of panel design than a factor of system setup. Exposure to direct sunlight, hours of sunlight, and amount of wind, clouds, or fog are all important factors that will impact on efficiency when setting up a system.

Solar panels are made from plastic or metal and are then glazed (covered in glass) or left unglazed. Obviously the glazed panel is heavier and more expensive; however, they absorb and retain more solar heat and are therefore more efficient (fewer panels are required to accomplish the same amount of solar heating).

OPEN LOOP SYSTEMS

The first system, shown in Fig. 3-6, is called an open loop system, meaning it is open to the pool water. This is the most common type you will encounter. Figure 3-7B shows a typical installation and plumbing diagram.

CLOSED LOOP SYSTEMS

The other type of system, found only in the most elaborate above-ground pool installations, is called a closed loop, where a separate pump circulates antifreeze through the solar panels. This liquid is heated in the panels and sent to coils inside a heat exchanger. The pool water is circulated through the heat exchanger, flowing around these coils so the heat from the coils is transferred to the water. These sys-

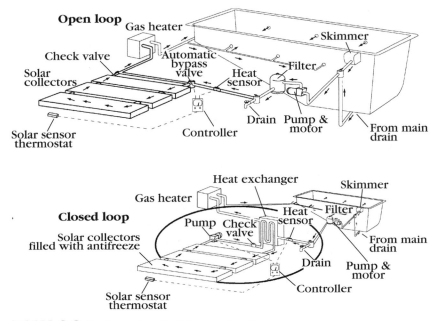

FIGURE 3-6 Typical solar plumbing and heating systems.

tems are used in cold climates or in desert climates, where it is hot by day but very cold by night, where water in an open loop system might freeze, expand, and crack the panels. Another advantage of the closed loop is that harsh chemicals in the pool water or hard, scaling water are not circulated into the panels with the potential to clog or corrode them, creating the need for expensive repairs.

Plumbing

Plumbing for solar heating is no different from other pool plumbing. It is located between the filter and the heater (Fig. 3-6) so water going to the solar panels is free of debris and is available for free solar heating before costly gas or electric heating by the system's mechanical heater.

Most solar heating systems for above-ground pools are controlled by manual three-port valves, but they can also be operated automatically. A thermostat on the solar panel determines the water temperature, and if it is warmer than the water coming out of the filter, a three-port motorized valve (called an automatic bypass valve) sends the water to the solar panels for heating and returns it to the plumbing

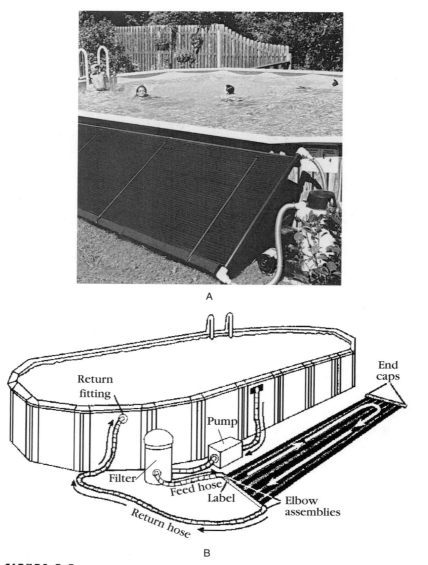

FIGURE 3-7   (A) Pool with solar panel unit. (B) Basic solar heating system layout. *SmartPool, Inc.*

that enters the heater. The heater thermostat senses the temperature of this solar heated water, and if it is still not as hot as desired, the heater will come on to heat it further before returning it to the pool.

Therefore, whether manual or automatic, a main component of solar heat plumbing is the three-port valve that either sends the water

QUICK START GUIDE: ADD SOLAR HEATING TO YOUR POOL

1. **Prep**
   - Unpack solar heating kit: panels, plumbing, connectors. Read the owner's manual.
   - Shut off pump and tape over the switch or breaker so that no one can turn it on before you finish.
   - Isolate equipment plumbing by closing valves at suction line (at skimmer and/or main drain connection before pump) and return line (at pool discharge outlet).

2. **Set Up**
   - Mount panels per owner's manual instructions on ground, deck, prefabricated rack, or roof.

3. **Plumb**
   - Cut pool equipment plumbing AFTER the filter but BEFORE the heater (if your system has one).
   - Plumb solar panel "intake" line to discharge pipe from filter; plumb solar panel "outlet" line to pool return line (or "intake" line of heater if you have one). Use shutoff valves at each location.

4. **Start Up**
   - Reopen pool plumbing valves (and valves on solar panel plumbing).
   - Start pump, purge air, and check for leaks. The solar panels are now in operation.

from the filter directly to the heater or sends it first to the solar panels and then the heater. A check valve is installed on the pipe that returns water from the solar panel to the heater to prevent water from entering this return line when the solar panels are not in use. This might instead be another three-port valve that performs the same function as the check valve but also ensures that when not in use, the solar panels will not drain out. This might not be important where solar panels are installed at or below the water level of the equipment and pool, but many installations of panels are on rooftops, high above the water level. Ball and check valves should be used so that the solar heating system can be completely and easily isolated from the circulation system (Fig. 3-8), allowing normal pool operation when repairing the solar panels.

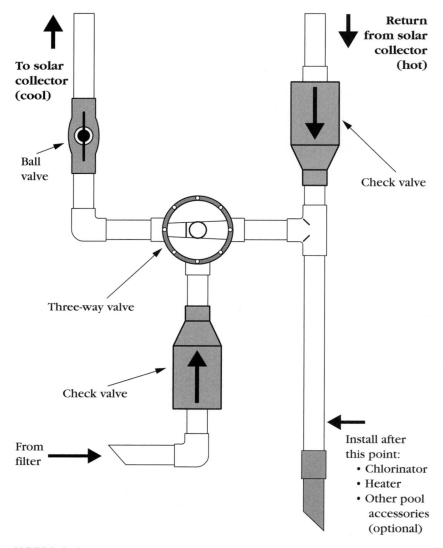

FIGURE 3-8 Solar system circulation plumbing.

To Solar or Not to Solar?

The decision to invest in a solar heating system will be based on the desired length of the "swimming season" and the desired swimming temperature. If nothing is done to prevent heat loss (or to add heat) to the pool, the water temperature will closely resemble the average air temperature. Therefore, if you want a swimming temperature above

70°F (21°C) and the average air temperature in your area meets that criterion only in the months of June through September, then that is your swimming season. It may take only a few hours per day of solar heating to raise the water temperature above 70°F in the months just before and after that period, thus easily doubling your swimming season.

To effectively heat with solar, regardless of the type of panel, the general rule of thumb for above-ground pools is that 35 percent of the surface area of the pool is the surface area of panels needed. For example, if the pool is 20 by 40 feet (6 by 12 meters), the surface area is 20 × 40 = 800 square feet × 0.35 = 280 square feet (26 square meters) of panels needed. Panels are sold in various convenient sizes ranging from 4 by 12 feet (1.2 by 3.7 meters) up to 8 by 25 feet (2.4 by 7.6 meters). Our example pool would need at least one of the largest panels, providing 200 square feet (18.6 square meters) of heating surface, somewhat less than the optimal size as calculated. A better option for this pool would be two panels of 8 by 20 feet (2.4 by 6 meters) plumbed together for a total of 320 square feet (30 square meters) of heating surface. When planning a solar heating installation, I always recommend slightly more than is absolutely needed because so many variables will conspire to reduce the effectiveness of any system once it's in operation.

Some say as little as 25 percent of the pool surface area can be used for these calculations, but I have found it is better to have a few more square feet of panels because you can't really have too many—but you can certainly have too few. Added panels will heat the water faster or, at least, more effectively on cloudy or windy days. For the few extra dollars, you will be happier in the long run. Figure 3-9 will assist with estimating probable efficiency and therefore overall sizing. Orientation due west, for example, will require solar panels equal to at least 45 percent of the pool's surface area. But the same installation oriented due south will require only 38 percent. These concepts are based on the northern hemisphere and will be exactly opposite in the southern hemisphere. The manufacturer can tell you the weight of each panel with water, so you can determine if the roof or deck has the space and weight-bearing capacity for the installation.

Next, a location must be found where the panels can face the sun. It might differ in your area, but generally the best position is facing south to obtain the most hours of sun per year—winter and summer. Another factor is prevailing wind. High winds can tear panels from the roof or

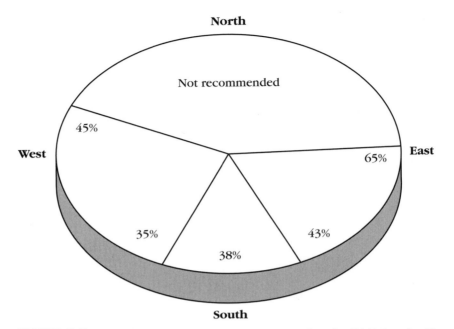

FIGURE 3-9 Solar panel sizing guide for above-ground pools. (1) Determine the area of the water surface of your pool. (2) Locate on the chart the direction your solar panels will face. (3) Multiply the percentage taken from the chart by the surface area of the pool to determine the total area of panels needed to effectively heat the pool. Example: The pool's water surface is 500 square feet and the panels will face due west; 45 percent of 500 = 225 square feet (21 square meters) of panel surface required.

create so much cooling that the panels will not be effective. Such concerns will also dictate how many panels you need to install.

Finally, to estimate the cost of the initial investment, as a general rule of thumb it will cost about $5 to $8 per square foot or 930 square centimeters for a solar heating system.

Knowing only these facts, you can determine if solar is likely to be practical. The cost of the system will be paid off with energy savings and tax benefits depending on how much gas or electricity would otherwise be used.

Installation

RATING: ADVANCED

If you decide to proceed, you might want to purchase one of the many fine solar heating packages from your pool supply house that includes panels, plumbing, controls, and instructions. You might want to hire a licensed carpenter (let him or her get the building permits and take the liability) to help with the installation and support of the panels if your installation is being mounted any place except on flat ground, while you complete the plumbing into the system.

Installation of a solar heating system will need to consider:

- Orientation, pitch, and location
- Size
- Hydraulics (if roof-mounted)
- Mounting
- Controls and automation
- Monitoring and isolation

We have already reviewed orientation. The angle of the panels ("pitch") as they sit on the roof or ground is also important, because the more the sun's rays strike the panels at a 90-degree angle, the more heat will be absorbed into the water. Therefore, a 20- to 30-degree pitch helps the efficiency of the system in winter when the sun tends to be on the horizon rather than directly overhead.

The existing pump will probably be adequate for circulation when adding solar equipment. For roof-mounted systems, you do, however, have to calculate the effect of the length of pipe and fittings as with any plumbing installation. Check with a pool professional for hydraulics calculations.

Plumbing is always arranged so that water flows from the bottom of the solar panels toward the top and no more than 200 square feet (18.6 square meters) in any one array. Both of these measures assure even flow of water through the panels. If more than 200 square feet of panels are needed, they can be plumbed as shown in Fig. 3-10.

As mentioned, there are numerous manufacturers and styles of panels and controls, far more than can be outlined here. Do some homework on what is available in your area. Just so you know what to look for, here are a few types:

- Rubber panels (like doormats) that nail directly to the roof or mount next to the pool (Fig. 3-7B). The entire system is rolled up into a small box and is easily installed in 30 minutes.

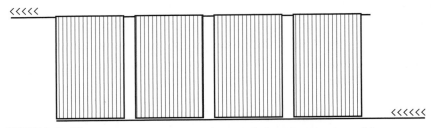

FIGURE 3-10 Water flows through solar panels from bottom toward top.

- Plastic panels (glazed or unglazed).

- Metal panels (glazed or unglazed).

- Thin, lightweight aluminum panels.

- Flexible plastic or metal hose that is coiled on the roof or built into a deck (the sun heating the deck in turn heats the solar coils).

Whatever the style, remember when planning an installation that long runs of pipe to and from the panels should be insulated so heat is not lost along the way. It is acceptable to lay solar panels on grass, but expect that part of the lawn to die. If you build a rack to mount the panels off the ground, make sure it isn't so close to the pool that children will be tempted to use it as a diving platform.

Although the panels themselves are fairly breakproof, choose a simple style that doesn't require replacement parts from manufacturers who might not be in business next year when you need to make repairs. Better yet, choose a manufacturer who has been around awhile.

Maintenance and Repair

RATING: EASY

Once installed, most homeowners and pool technicians tend to forget about the solar heating system. Inspection every two or three months should be made to check for leaks. Leaks can easily occur because of the extremes of hot and cold temperatures that cause the panel materials to expand and contract. Leak repair depends on the type of material in the panel or plumbing, and each manufacturer makes leak repair kits with instructions. The plumbing to and from the panels can be repaired as needed using the techniques outlined in Chap. 2 on basic plumbing.

TRICKS OF THE TRADE: SOLAR HEATING TROUBLESHOOTING

Now that we have covered the most common problems, here is a troubleshooting guide that may assist in quickly identifying and repairing solar heating difficulties:

Pool not as warm as it should be

- Panels too small or incorrectly oriented
- Circulation through panels not long enough each day
- Circulation at wrong time of day (if water goes through the panels at night, the water may be cooling instead of heating)
- Auto controls not working properly
- Water flowing through panels too fast
- Panels dirty

Air bubbles at pool return lines only when solar is operating

- Check for clean filter
- Vacuum relief valve not operating properly or clogged

Some panels warm to the touch, others cool

- Check for circulation problems
- Check that each array is no more than 200 square feet
- Check flow rate
- Check for any valves between panels

Leaks in panels or plumbing

- Check water chemistry
- Check that panels and plumbing are secure

If just one tube is leaking, use a razor knife to cut if off of each manifold. Use a stainless steel sheet metal screw (typically no. 10) with Teflon tape on the threads to seal the resulting hole in each manifold.

The second common problem is dirty panels. Dirt prevents the panels from absorbing heat and can cut efficiency by as much as 50 percent. A pool technician can make good profits by charging customers for regular solar panel cleaning, requiring no more than soap and water.

By the way, don't be concerned that plastic panels are not functioning properly because they are cool to the touch. When water is circulating

FAQS: ADVANCED PLUMBING SYSTEMS

Should I add valves to my plumbing system?

- **Yes. Shut-off valves at the suction and discharge points of your plumbing (as close to the pool as possible) will allow you to isolate plumbing and equipment for maintenance and repair. They also help if you discover a leak and need to determine if it is in the pool itself or the plumbing and equipment.**

Will a solar heating system pay for itself?

- **If you are heating your pool with gas or electricity, a solar heating system will pay for itself in 2 to 5 years, depending on your use of the pool and the heating fuel. If it is your sole source of pool heat, it pays for itself by significantly extending the swimming season.**

Must solar panels be mounted on a rooftop to be effective?

- **No. Most solar panels for above-ground pools are set up on the ground near the pool. The key to their efficiency is angle and exposure to the sun, regardless of where they are mounted. Wind is another important factor—panels on a roof might be cooled sooner than panels set on the ground in a more wind-protected area.**

through the panel, it will transfer the heat to the water and away from the plastic surfaces. If the solar panel has been shut off for some time, the panel will be hot to the touch until water begins to circulate. Be careful to warn swimmers to stay away from the return water discharge port in the pool when reopening the solar heating system because the water that has been sitting in the panel can be scalding.

Finally, panels can become clogged with scale from hard pool water and chemicals. Poor circulation is the tip-off, and the solution is either to replace the panel or, if possible, to disassemble the panels from each end, exposing the pipes of the panel that actually carry the water, and reaming these out with special brushes attached to your power drill. Again, how you make this repair depends on the maker of the panel and its style. The maker should provide instructions and special tools for this procedure. As with leak repair or cleaning of solar panels, reaming is simple to perform using techniques and skills learned elsewhere in this book.

CHAPTER 4

Pumps and Motors

Let's begin by eliminating a common error in terminology. The pump and motor are two different elements of the water circulation system, not the same thing, not interchangeable. The *motor* is the device that converts electrical energy into mechanical energy. It powers the *pump*, which is the device that actually causes the water to move. One is not much use without the other.

While built of various metals or plastics, all pump and motor combinations are composed of essentially the same components. If you understand the basic concept and components, you can find your way around almost any pump or motor. Before discussing the components of a pump and motor, let's understand the concept of what they do and how they work.

How They Work

Pool pumps are classified as *centrifugal* pumps. That is, they accomplish their task of moving water thanks to the principle of centrifugal force. To imagine this concept, hold a bucket with some water in it at the end of your arm and spin it around in a big sweeping circle (Fig. 4-1).

Centrifugal force is what keeps the water in the bucket as you spin it. If you poke a hole in the bucket and spin it again, that same force, pushing the water to the bottom of the bucket, sends it shooting out the

FIGURE 4-1 Centrifugal force. *Sta-Rite Industries.*

hole. If you spin the bucket faster, the water shoots out with more force. Obviously, if you make a larger hole, more water will shoot out as you spin it around.

The pump operates the same way (Fig. 4-2). The impeller in the pump spins, shooting water out of it. As the water escapes, a vacuum is created that demands more water to equalize this force. Water is pulled from the pool and sent on its way through the circulation plumbing. Just as various designs in your swinging bucket and its hole determine the amount of water and how fast it escapes, so too the various designs of impellers, diffusers, and volutes determine the same features in a pool pump. This is discussed in more detail in later sections.

Pumps used for pools are *self-priming*; that is, they expel the air inside upon start-up, creating a vacuum that starts suction. Once water is flowing through the pump, if you close a valve on the outflow side of the pump, restricting all flow, maximum possible pressure is created. However, unlike a piston or gear pump, there is no destructive force created—the impeller simply spins in the liquid indefinitely.

Intake port   Lid   Discharge port

Strainer pot   Strainer basket   Seal   Volute   Impeller   Seal plate

FIGURE 4-2 Typical pool pump (cutaway view).

Let's examine the components of the pump and motor partnership (Figs. 4-3 and 4-4).

Strainer Pot and Basket

The plumbing from the pool skimmer runs to the inlet port of the pump, which is usually female threaded for easy plumbing, although some designs are male and female threaded.

Water flows into a chamber, called the *strainer pot* or *hair and lint trap*, which holds a basket (generally 4 to 6 inches or 10 to 15 centimeters in diameter and 5 to 9 inches or 13 to 23 centimeters deep) of plastic mesh that permits passage of water but traps small debris. Some baskets simply rest in the pot, others twist-lock in place.

Most have handles to make them easier to remove, although I have yet to see a design where the handle is not so flimsy that it doesn't

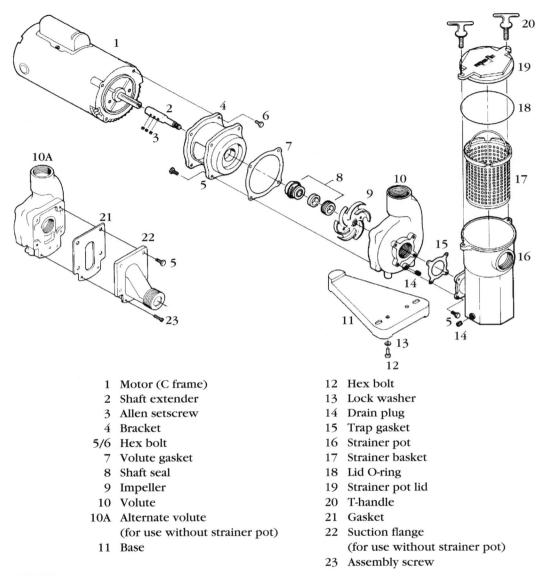

1	Motor (C frame)	12	Hex bolt
2	Shaft extender	13	Lock washer
3	Allen setscrew	14	Drain plug
4	Bracket	15	Trap gasket
5/6	Hex bolt	16	Strainer pot
7	Volute gasket	17	Strainer basket
8	Shaft seal	18	Lid O-ring
9	Impeller	19	Strainer pot lid
10	Volute	20	T-handle
10A	Alternate volute	21	Gasket
	(for use without strainer pot)	22	Suction flange
11	Base		(for use without strainer pot)
		23	Assembly screw

FIGURE 4-3 Typical bronze pump and motor, exploded view. *Aqua-Flo, Inc.*

break off the second or third time you use it. The strainer basket is similar to the skimmer basket which traps larger debris.

The strainer pot is a separate component in some pumps that bolts to the volute with a gasket or O-ring in between to prevent leaks. Sometimes the pot includes a male threaded port that screws into a

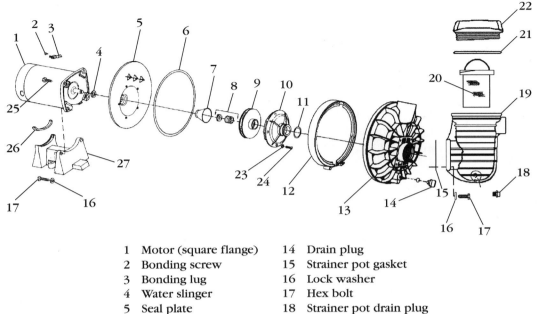

FIGURE 4-4 Typical Noryl plastic pump, exploded view. *Sta-Rite Industries.*

1	Motor (square flange)	14	Drain plug
2	Bonding screw	15	Strainer pot gasket
3	Bonding lug	16	Lock washer
4	Water slinger	17	Hex bolt
5	Seal plate	18	Strainer pot drain plug
6	Seal plate O-ring	19	Strainer pot
7	Seal insert	20	Strainer basket
8	Shaft seal	21	Lid O-ring
9	Impeller	22	Strainer pot lid
10	Diffuser	23	Star washer
11	Diffuser O-ring	24	Assembly machine screw
12	Clamp	25	Assembly allen-head bolt
13	Volute	26	Motor mount pad
		27	Motor mount

female threaded port in the volute. In some pumps, it is a component molded together with the volute as one piece (Fig. 4-2).

To clean out the strainer basket, an access is provided. The strainer cover is often made of clear plastic so you can see if the basket needs emptying. It is held in place by two bolts that have a T top (Fig. 4-3) or plastic handle (Fig. 4-4) for easy gripping and turning. Some pumps have a metal clamp that fits around the edge of the cover and strainer pot. These are tightened with a bolt and nut combination. On others the cover is male threaded (Fig. 4-4) and screws into the female threads of the pot.

In all styles of pot, the strainer cover has an O-ring that seats between it and the lip of the strainer pot, preventing suction leaks. If this O-ring fails, the pump sucks air through this leaking area instead of pulling water from the pool or spa.

Notice in Figs. 4-3 and 4-4 that the pot has a small threaded plug that screws into the bottom. This plug is designed to allow complete drainage of the pot when winterizing the pump. It is made of a weaker material than the pot (on metal pots, for example, the plug is made of plastic, soft lead, or brass). If the water in the pot freezes, this sacrificial plug pops out as the freezing water expands, relieving the pressure in the pot. Otherwise, of course, the pot itself will crack.

Volute

The volute is the chamber in which the impeller spins. Combined with the impeller, the volute forces water out of the pump and into the plumbing that takes the water to the filter. The outlet port is usually female threaded for easy plumbing. When the impeller (Fig. 4-5) moves water, it sucks it from the strainer pot.

The resulting vacuum in the pot is compensated for by water filling the void. The rushing water is contained by the volute which directs it out of the pump. Therefore the pot can be considered a vacuum chamber and the volute a pressure chamber.

The impeller by itself will move the water, but it cannot create a strong vacuum by itself to make the water flow begin. The area immediately around the impeller must be limited to eliminate air and help start the water flow. A diffuser (Fig. 4-4) and/or closed-face impeller help this process, but in many pump designs, the volute serves this purpose.

Figure 4-3 shows how the volute closely encircles the impeller. Figure 4-4 shows a design with a separate diffuser that houses the impeller. In some designs, the inside of the volute is ribbed to improve the flow efficiency.

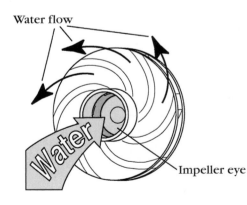

Water flow

Impeller eye

FIGURE 4-5 Water's-eye view of the impeller. *Sta-Rite Industries.*

Impeller

The impeller is the ribbed disk (the curved ribs are called vanes and the disk is called a shroud) that spins inside the volute. As water enters the center or eye of the impeller (Fig. 4-5), it is forced by the vanes to the outside edge of the disk, just like our spinning bucket example. As the water is moved to the edge, there is a resulting drop in pressure at the eye, creating a vacuum that is the suction of the pump. The amount of suction is determined by the design of the impeller and pump components and the strength of the motor that spins the impeller.

There are essentially two types of impellers: closed-face (Fig. 4-5) and semiopen-face (Fig. 4-6). Some publications call the semiopen-face impeller an open face. This is not accurate, because for the pump to be self-priming, which most pool pumps are, it needs a disk (shroud) on the front face as well. Therefore, although a particular impeller itself has no front shroud and might be called open when it stands alone, it does in fact make use of some sort of front shroud, either a diffuser located closely around the impeller or the interior surface of the volute.

In Fig. 4-6, you get another water's-eye view of a volute and impeller. The side view shows how close the "open-face" impeller is to the interior side of the volute, effectively forming a front shroud. The clearance between the volute interior and impeller face is critical. Too far away and there will be insufficient pressure created in the volute, causing weak or no suction. Too close and the impeller might rub against the volute or jam if small debris lodges between the two. As discussed later in the section on the shaft, the semiopen impeller pump can be adjusted for optimum efficiency of the impeller.

In the closed-face impeller, as the name implies, the vanes of the impeller are covered in both front and back. Water flows into the hole in the center and is forced out at the end of each vane along the edge of the impeller. This type of impeller, especially in connection with a diverter, is extremely efficient at moving water.

If the closed-face impeller is so much more efficient and requires no shaft extender or adjustment, why have the semiopen-face impeller designs survived? Because the downside of the closed face is that small stones, pine needles, and other fine debris can get past both the skimmer and strainer baskets and clog the closed vanes, slowing or completely shutting off water movement. The semiopen-face design allows this small debris to pass (actually it is usually pulverized) to the filter, although it is not impossible for a heavy volume of small debris to clog the open vanes as well.

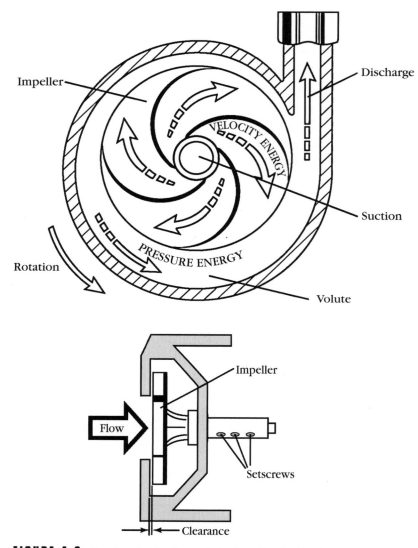

FIGURE 4-6 Front and side views of the impeller inside the volute. *Sta-Rite Industries.*

One of the chief culprits in clogging of both semiopen and closed impeller designs is DE (diatomaceous earth). As is described in the chapter on filters, this white, powdery material is used to precoat some designs of filter grids. If introduced too quickly, DE will clog any restricted area—plumbing elbows, strainer baskets, and impeller vanes.

Most impellers on pool pumps have a female threaded hole on the center back side (the side facing away from the water source) that screws onto the male threaded end of the motor shaft (or shaft extender). The rotation of the shaft is just like a bolt being threaded into a nut. As the shaft turns, it tightens the impeller on itself. Others, like the old Purex AH-8 models, are fitted with setscrews that clamp the impeller onto the end of the motor shaft.

Impellers are rated by horsepower to match the motor horsepower that is used.

Seal Plate and Adapter Bracket

The volute is the pressure chamber in which the impeller spins to create suction. If this were all one piece, there would be no way to remove the impeller or to access the seal. Therefore this chamber is divided into two sections. The actual curved housing is called the volute, while its other half is called the seal plate or adapter bracket.

The seal plate (Fig. 4-4, item 5) is joined to the volute with a clamp. An O-ring between them makes this joint watertight. The motor is bolted directly onto this type of seal plate. In other designs, the seal plate is molded together with an adapter bracket that supports the motor (Fig. 4-3, item 4) and bolts to the volute, with a paper or rubber gasket between them to create a watertight joint. In yet another style, the pump sections are joined with a threaded union type of clamp, like the lid of a jar. This allows disassembly by hand.

In both cases, the shaft of the motor passes through a hole in the center of the seal plate and the impeller is attached, threaded onto the shaft. The bracket allows access to the shaft extender (see following section) for adjusting the clearance between the impeller and volute (Fig. 4-6). The pump design shown in Fig. 4-4 is a closed face and needs no such adjustment, so the shaft need not be exposed.

Shaft and Shaft Extender

The shaft of the motor is the part that turns the impeller, creating water flow. Figure 4-3 shows a motor with shaft. In this design of pump, the impeller needs to be adjusted in relation to the volute, so a shaft extender has been created. The extender, made of brass or bronze, slides

TOUGH NUT TO CRACK

Note, in Fig. 4-4, that the clamp (item 12) that joins the volute and seal plate is made tight by a bolt and nut. Never assemble the pump with this bolt underneath the pump. When the pump is bolted to the deck, it makes it a knuckle-busting, cussword of a job to remove the clamp. Moreover, if the pump leaks at all, the bolt gets and stays wet, causing it to rust, making unscrewing it even tougher, or the bolt breaks altogether. If you come across one, move the clamp bolt to the top of the pump before installation.

over the motor shaft and is secured by three allen-head setscrews (Fig. 4-6). The male threaded end of the shaft extender then fits through the seal plate and the impeller is screwed into place.

Note that the extender is round, but a flat area has been created on two sides. In this way, a ¾-inch (19-millimeter) box wrench can be used to prevent the extender from spinning when performing maintenance. Note also that some designs require an O-ring near the threads of the extender to ensure a watertight seal. Other designs rely only on the pump seal. Once assembled, the clearance between the impeller and volute can be adjusted as shown in Fig. 4-6 (side view).

The shaft of the motor in this example (Fig. 4-3) is called a keyed shaft. This means the cylindrical shaft has a groove running the length of the shaft to accept the setscrews. In this way, when the setscrews are in place, they prevent the motor shaft from slipping or skidding inside the extender, thus failing to turn the impeller. Figure 4-4 shows a pump style that requires no shaft extension. The motor shaft, already engineered to the exact length required, has a threaded end to accept the impeller.

The shaft should never be in contact with electric current, but water is a great conductor and wet conditions around pool equipment can circumvent the best of designs. Because of this, most motor shafts today are designed with a special internal sleeve to insulate the electricity in the motor from the water in the pump.

Seal

Obviously if the shaft passed through the large hole of the seal plate without some kind of sealing, the pump would leak water profusely. If the hole was made small and tight, perhaps of tight-fitting rubber, the high-speed spinning of the shaft would create friction and burn up the components or the shaft would bind up and not turn at all. Some clever engineer devised a solution to this problem called a *seal*.

The seal allows the shaft to turn freely while keeping the water from leaking out of the pump. In Fig. 4-3, the seal (item 8) is in two parts. The right half of the seal is composed of a rubber gasket or O-ring around a ceramic ring. This assembly fits into a groove in the back of the impeller. The ceramic ring can withstand the heat created by friction. The left half fits into a groove in the seal plate and is composed of a metal bushing containing a spring. A heat-resistant graphite facing material is added to the end of the spring that faces the ceramic ring in the other half.

The tight fit of the seal halves prevents water from leaking out of the pump. The spring puts pressure on the two halves to prevent them from leaking. As the shaft turns, these two halves spin against each other but do not burn up because their materials are heat-resistant and the entire seal is cooled by the water around it. Therefore, if the pump is allowed to run dry, the seal is the first component to overheat and fail.

The pump design in Fig. 4-4 includes an additional seal housing or insert. If the pump runs dry, the heat build-up not only melts the seal, but also the plastic seal plate in which it is mounted. The inset helps isolate the seal plate from this heat and the cone-shaped unit diffuses heat.

Pump makers are always improving the heat dissipation (heat sink) capabilities of their pumps so that dry operation will result in little or no damage to the seal or pump components. Still, pumps are not designed to run without water for more than a few minutes while priming.

Motor

Before reading this section, a basic knowledge of electricity is helpful, so you might want to review a book on basic electricity. After all, the motor is the device that converts electricity into mechanical power.

Motors, like the pumps they drive, are rated by horsepower, usually in pool and spa work as $\frac{1}{2}$, $\frac{3}{4}$, 1.0, 1.5, and 2.0 horsepower. Commercial installations might use higher rated systems, but these are the most common.

As shown in Fig. 4-7, electricity flows through the motor windings, which are thin

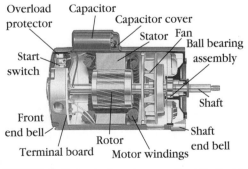

FIGURE 4-7 **Typical pool motor.** *Franklin Electric.*

strands of coiled copper or aluminum wire. The windings magnetize the iron stator. If you paid attention in your first grade science class, you recall that opposite poles of a magnet attract each other, but like poles repel. Using this concept, the rotor spins, turning the shaft. Some designs employ one set of windings for the start-up phase where greater turning power (torque) is needed and another set for normal running.

The shaft rides on ball bearings at each end. A built-in fan cools the windings because some of the electrical energy is lost as heat. The caps on each end of the motor housing are called end bells. A starting switch is mounted on one end with a small removable panel for electrical connection and maintenance access.

This is where you will also find the thermal overload protector. This heat-sensitive switch is like a small circuit breaker. If the internal temperature gets too hot, it shuts off the flow of electricity to the motor to prevent greater damage. As this protector cools, it automatically restarts the motor, but if the unit overheats again, it will continue to cycle on and off until the problem is solved or the protector burns out.

It takes a great deal of electricity to start a motor but far less to keep it going (in fact, about five to six times less). The capacitor, as the name implies, has a capacity to store an electrical charge. The capacitor is discharged to give the motor enough of a jolt to start, then it is able to run on a lower amount of electricity. Without the capacitor, the motor would need to be served by very heavy wiring and high-amp circuit breakers to carry the starting amps. The start-up amperage of a motor is about twice that of its running amperage. The capacitor is located in a separate housing mounted atop the motor housing (as in Fig. 4-7) or inside the front end bell.

Some motors are designed to operate at two speeds. For example, some spas operate at high speed for jet action, but lower speed for circulation and heating. In pool and spa work, the normal rotation speed is 3450 revolutions per minute (rpm) and the low speed is 1750 rpm.

Energy-Efficient Motors

Energy-efficient motors have heavier wire in the windings to lower the electricity wasted from heat loss.

A good way to compare energy efficiency between two motors is to compare the gallons pumped to kilowatts used. Let's say one pump produces a flow rate of 50 gallons (189 liters) per minute, which is

3000 gallons (11,355 liters) per hour. Divide that by the kilowatts used per hour. The higher the resulting number, the more efficient is the pump and motor.

By the way, kilowattage is determined by multiplying amps by voltage. Let's say the unit runs at 9 amps at 220 volts—9 × 220 = 1980 watts. Kilowatts (meaning 1000 watts) would then be 1980 divided by 1000. So this unit uses 1.98 kilowatts each hour. The 3000 gallons divided by 1.98 equals a rating of 1515 gallons (5734 liters) pumped for every kilowatt used. You can now make similar calculations for other pump and motor units for comparison.

Voltage

I have been discussing typical 110/220-volt motors. In fact, most motors are designed to be connected to either power source. By changing a wire or two internally, you determine which voltage is used. The instructions for this conversion are printed on the motor housing or inside the access cover.

If your motor is wired for 220 volts and you feed it 110 volts, it will run slowly or not start. If your motor is wired for 110 volts and you feed it 220 volts, the thermal overload protector should overheat and cut the circuit. Higher horsepower motors might run on three-phase current. You don't want to fool with that. Call an electrician.

Housing Design

The housing of the motor is designed to adapt to various pump designs. Figure 4-3 shows a motor called a *C frame*, because the face of the motor resembles a C. All this means is that it will fit certain kinds of pumps. Figure 4-4 shows a motor called a square *flange*, for equally obvious reasons. There are other types, such as the 48, uniseal flange, and those designed for automatic pool cleaner booster pumps. When replacing a motor, you need to buy the proper housing type.

Turnover Rate

So now you know what a pump and motor are, and, if you must replace either, in most cases you will assume the original designer or builder used the correct size pump and motor for the job and make your replacement with the same size. Or will you?

What if the original equipment was too small or too large? What if the plumbing has been repaired (which might have added or deleted pipe and fittings) or equipment has been added or deleted, thus changing the system and requiring the pump to work more (or less)? What if the identifying rating plates have been removed or are so weatherworn that you can't tell what size the existing equipment is? Finally, what if it is a brand new installation? How do you decide what is the right pump and motor for the job?

Well, I'm glad you asked all of those intelligent questions. The answer is based on turnover rates.

The *turnover rate* of a body of water is how long it takes to run all the water through the system. It is desirable for the water to completely circulate through the filter one to two times per day, but a better rule of thumb is that a pool must turn over in 6 hours.

Let's say you've calculated the volume of water in the pool (see Chap. 1) as 18,000 gallons (68,000 liters).

$$18{,}000 \text{ gal} \div 6 \text{ hours} = 3000 \text{ gal } (11{,}333 \text{ L}) \text{ per hour}$$

$$3000 \text{ gph} \div 60 \text{ min} = 50 \text{ gpm } (188 \text{ lpm})$$

Therefore, you need a pump capable of delivering a flow rate of 50 gpm (188 lpm). The manufacturer's specifications will tell us which pump can do this.

Turnover rate is also a factor of the pool's hydraulics and the horsepower rating of the pump and motor. *Hydraulics* is the study of water flow and the factors affecting that flow, such as pipe diameter, length, and other physical factors that constrain the movement of water. In most above-ground pool systems, the equipment is located close to the pool and the pipes are short and visible, so complex calculations of the hydraulics are unnecessary when replacing pool equipment. If your above-ground pool has remotely located equipment with long, buried pipes to the pool, you may want to consult the hydraulics section of *The Ultimate Pool Maintenance Manual* when replacing a pump or making major changes to the plumbing.

Maintenance and Repairs

Since the pump and motor are the heart of the system, if they fail or don't perform efficiently, the abilities of the other components won't much matter. Keeping the motor in good working order is a matter of keeping it dry and cool. The best detection tool for motor problems is your ears. Laboring motors or those with bad bearings will let you know.

Keeping the pump in good order is also a matter of sight. Seeing leaks tips you that the pump needs attention. If the motor needs to stay dry, but problems with pumps most often result in leaks, the potential for pump and motor breakdown is high. Therefore, keeping an eye and ear on your pump and motor will pay dividends in a pool that keeps running.

The basic repairs and maintenance of the pump and motor unit, starting from the front, the first place the water encounters, are discussed in the following paragraphs.

Pump and Motor Health Checklist

Look

- Motor dry
- Vents free of leaves or other debris
- No pump leaks
- Strainer pot clean

Listen

- Steady, normal hum
- No laboring, cavitating, or grinding noises

Feel

- Motor warm, but not hot
- No major vibration

Strainer Pots

RATING: EASY

Clean out the strainer basket often. Even small amounts of hair or debris can clog the fine mesh of the basket and substantially reduce flow. To be honest, this job is a pain in the valve seat. You have to shut down the system, struggle with tight cover bolts or clamps, clean out hair and filth from the basket, put the basket back, find a water source to fill the pot so the pump will reprime easily, check the

---

TOOLS OF THE TRADE: PUMPS AND MOTORS

- Flat blade screwdriver
- Phillips screwdriver
- Allen-head wrench set
- Open end/box wrench set
- Hacksaw
- $5/16$-inch (8-millimeter) nut driver
- Impeller wrenches
- Teflon tape
- Silicone lube
- Needle-nose pliers
- Hammer
- Seal driver
- Emery cloth or fine sandpaper
- Impeller gauge
- Tap and die set
- Razor knife

O-ring, replace the cover, tighten the bolts or clamps, restart the system, and most often, reprime . . . whew!

But this, along with keeping a clean skimmer basket, are the two most simple and important elements to keep a pool clean and the other components working. If the water can't flow adequately, it can't filter or heat adequately either. It will turn cloudy, allow algae growth, and make vacuuming difficult.

The only other problems you might encounter at the strainer pot are broken baskets or a crack in the pot itself. If the basket is cracked it will soon break, so replace it. If allowed to operate with a hole in it, the basket will permit large debris and hair to clog the impeller or the plumbing between the equipment components.

Cracks might develop in the pot itself, especially if you live where it gets cold enough to freeze the water in the pot. Again, the only remedy is replacement. Follow the directions described in the following paragraphs for changing a gasket, because it requires the same disassembly and assembly techniques.

Gaskets and O-Rings

RATING: EASY

Most problems occur in strainer pots when the pump is operated dry. The air heats in the volute as the impeller turns without water to cool it. The strainer basket will melt; the pot cover, if plastic, will warp; and the O-ring will melt or deform. Usually, replacement of the overheated parts solves the problem.

GASKETS

When gaskets leak, or, in extreme cases, if the strainer pot itself must be replaced (Fig. 4-3, items 15 and 16), the replacement process is the same. Remove the strainer pot [take out the four bolts, usually using a ½- or ⁹⁄₁₆-inch (13- or 14-millimeter) box wrench]. Clean out the old gasket thoroughly. Failure to do this will leave gaps in the new gasket that will eventually leak. Reassemble the new gasket and strainer pot the same way the old one came off. Tighten the bolts evenly (so the new gasket compresses evenly) by gently securing one bolt, then the one opposite, then the last two. Continue tightening in this crisscross pattern until each bolt is hand tight. When dealing with plastic pumps, do not overtighten because the bolt will crack the pump components or strip out the female side.

Sometimes the bolts are designed to go through the opening in the pot and volute and are tightened with a nut and lock washer on the other side. Still, do not overtighten, because you will crack the pump components. The key to this simple procedure, as with virtually all other mechanical repair, is to carefully observe how the item comes apart. It will go back together the same way.

O-RINGS

When removing and replacing the strainer pot cover (Fig. 4-3, item 19), be sure the O-ring (Fig. 4-3, item 18) and the top of the strainer pot are clean because debris can cause gaps in the seal. Sometimes these O-rings become too compressed or dried out and brittle and cannot seal the cover to the pot. In this case, replace the O-ring.

Pump and/or Motor Removal and Reinstallation

RATING: ADVANCED

Sometimes it is necessary to remove an entire pump and motor unit to take it apart or complete a repair. If the pump is damaged beyond your ability to repair it, you might want to take the entire unit to a motor repair shop. They can rebuild it as needed, and you can reinstall it at the job site. Your local pool and spa supply house can recommend a rebuilder, or consult the phone book.

Generally, to remove the pump and motor as a unit you will need to cut the plumbing on the suction and return side of the pump. Cut the pipe (Fig. 4-8) with enough remaining on each side of the cut

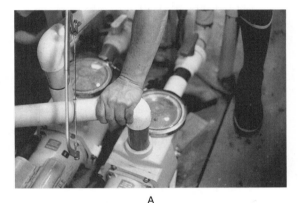

A

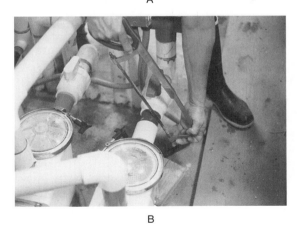

B

FIGURE 4-8 Pump and motor removal and installation: (A) step 1; (B) step 2.

TRICKS OF THE TRADE: O-RING EMERGENCIES

- If no replacement is available, try turning the O-ring over. Sometimes the rubber is more flexible on the side facing the cover. Be careful to remove the O-ring gently. Too much stress will cause the rubber to stretch, making it too large to return to the groove in the cover.

- If the O-ring stretches, try soaking it in ice water for a few minutes to shrink it.

- Coat the O-ring liberally with silicone lube. This can take up some slack and complete the seal if the O-ring is not too worn out.

- Another emergency trick is to put Teflon tape around the O-ring to give it more bulk and make it seal. If you use this trick, be sure to wind the tape evenly and tightly around the O-ring, so loose or excess tape does not cause an even worse seal. If the O-ring has actually broken, it will almost always leak at that spot; however, I have used the Teflon tape trick successfully in these cases for a temporary repair when a new O-ring was not immediately available.

TRICKS OF THE TRADE: SILICONE

- Remember to use only nonhardening silicone lube on all pool and spa work. Vaseline or other lubricants are made of petroleum, which eats away some plastics and papers.

- Get silicone lube at your supply house or any scuba diving shop—it is used for scuba equipment repairs for the same reasons.

- Before reassembly, coat pump halves with silicone where they contact gaskets or O-rings. This helps to fill any tiny gaps that might still be present, especially on older pumps.

to replumb it later. Ideally, a few inches (centimeters) on each side allows you to use a slip coupling to reglue the unit in place later (see the chapter on basic plumbing).

When installing or reinstalling the plumbing between the pump and filter take a look at the equipment area. Keep bends and turns to a minimum. Remember, each turn creates resistance in the system. Also, don't locate the pump close to the base of the filter. When you open the filter for cleaning, water is sure to flood the motor. Lastly, try to keep motors at least 6 inches (15 centimeters) off the ground. The bracket of the pump does this in part, but heavy rains or flooding from broken pipes and filter cleanings can flood the motor if it is too close to the ground.

When removing a pump and motor unit you have the opportunity to reinstall it on a raised surface for a greater margin of error. If you do this by adding a mounting block, don't use wood (it will deteriorate over time). A thick rubber mounting pad or large brick will work, but be sure to follow local building codes regarding bolting these to the deck and bolting the pump to the mounting material. All pumps should be bolted to the deck, but you will find many that are not. This is a good time to remedy such oversights.

The other component of pump and motor removal is the electrical connection. You have already turned off the breaker (right?). Now remove the access cover (Fig. 4-9A) to the switch plate area of the motor, near the hole where the conduit enters the motor. Remove the three wires inside the motor and unscrew the conduit connector (Fig. 4-9B) from the motor housing. Now you can pull the conduit and wiring away from the motor and the entire pump and motor should be free.

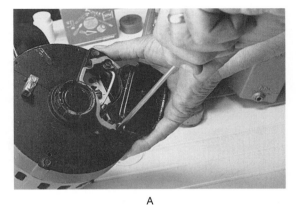

A

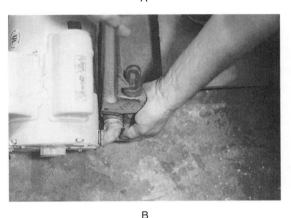

B

FIGURE 4-9 Motor electrical connections: (A) step 1; (B) step 2.

There might be an additional bonding wire (an insulated or bare copper wire that bonds/grounds all of the equipment together to a grounding system). This is easily removed by loosening the screw or clamp that holds it in place.

New Installation

RATING: ADVANCED

If you're lucky enough to install the pump and motor for the first time, do it right. I can't tell you how many installations look like they were done

SAFETY FIRST

Tape off the ends of the wires, even though the breaker is turned off, and put tape over the breaker switch itself. Leave a note on the breaker box to yourself, family members, or the customer to be sure no one accidentally turns the breaker back on while the pump and motor is away for service.

by someone who hated service technicians with everything plumbed together so tightly that later repairs were impossible knuckle-busters. You already know the plumbing and electrical techniques as discussed previously, so here are some tips on preparing for new installations.

Position the pump as close to the pool and as near to water level as possible so it doesn't have to work so hard. Mount the unit on a solid, vibration-free base, not wood (which rots). Make sure there is adequate drainage in the area so that when it rains or if a pipe breaks the motor won't be drowned. Bolt or strap down the pump as required by local code.

Plumb in both suction and return lines with as few twists and bends as possible, to minimize resistance. A gate valve on both sides is advisable to isolate the pump when cleaning other components. A check valve is essential if the unit is well above water level. Plumb the unit far enough away from the filter that it won't get soaked when you take the filter apart.

Replacing a Pump or Motor

RATING: ADVANCED

Having learned how to remove and break down a pump and motor in the previous sections of this chapter, replacing any of the components is simply a matter of disassembling the pump down to the component that needs replacement, getting a replacement part, and reassembling the unit. Of course, if the entire pump and motor is to be replaced, you purchase the replacement as a unit and plumb it in as previously described.

Sometimes the motor will trip the circuit breaker when you try to start it. If this happens it is usually because there is something wrong with the motor; however, it could be a bad breaker or one that is simply undersized for the job and has finally worn out. The section on motor troubleshooting (later) deals with checking the wiring, circuits, and breakers. However, to replace the motor depicted in Fig. 4-3, you follow the procedure of Fig. 4-10 as follows:

1. **Electrical** You can access the electrical connections through the switchplate cover in the front end bell (Fig. 4-9A). BE SURE THAT THE ELECTRICITY IS TURNED OFF! Turn off any equipment switch

or timer, and, for extra insurance, switch off the circuit breaker too.

2. **Disassembly** Remove the motor from the shaft extender by removing the allen-head setscrews (Fig. 4-10A) and pulling the extender off the motor shaft. Sometimes this might need persuasion. Use your large flat-blade screwdriver to pry the extender away from the motor body. Sometimes corrosion will eat away at the setscrews and extender—if it is too tough to remove, replace it (they're only a few bucks).

3. **Preparation** Before sliding the shaft extender on the new motor, clean the motor shaft with a fine emery cloth such as you might have in your copper pipe solder kit. Apply a light coat of silicone lube to the shaft—no, the silicone won't make the extender slip loosely on the shaft. When you put the extender on the motor shaft, the setscrews go into a groove that runs along the shaft. This groove allows the screws to grip and not slide around the shaft.

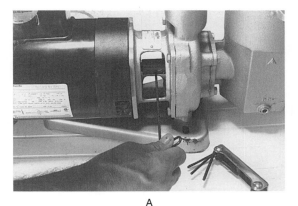

A

B

FIGURE 4-10 Replacing a motor: (A) step 1; (B) step 2.

4. **Reassembly** Secure the shaft extender by tightening the allen-head setscrews, but before doing so, be sure the impeller is properly positioned. The spring in the seal will push the impeller up against the inside of the volute when you loosen the setscrews, so now you must pull it back before securing the shaft extender. Insert your flat-bladed screwdriver into the neck of the shaft extender (where it passes into the seal) and gently pry it back toward the motor, exerting pressure against the spring in the seal: then tighten the setscrews. If you pry the extender too far back, the impeller will rub against the seal plate. Pry the extender all the way back, then let up ⅛ inch (3 millimeters) or so to obtain the right setting. When you restart the pump, if you hear any scraping or if the impeller won't

turn, you'll need to repeat this adjustment step. Reconnect the electrical connections.

Replacement of the motor for the Sta-Rite unit in Fig. 4-4 is the same process, but there is no shaft extender or adjustment to make when reassembling. All other pump and motor designs are variations on these themes and will be obvious once you have mastered these few steps.

Troubleshooting Motors

RATING: EASY

The first and most common motor problem is water. Motors get soaked in heavy rain, when you take the lid off the filter for cleaning, when a pipe breaks, or when you look at it wrong. In all cases, dry the motor and give it 24 hours to air dry before starting it up—moisture on the windings will short them out and short out your warranty as well. The basic problems beyond this are as follows.

MOTOR WON'T START

Check the electrical supply and breaker panel, and look for any loose connection of the wires to the motor. Sometimes one of the electrical supply wires connected to the motor switch plate becomes dirty. Dirt creates resistance that creates heat which ultimately melts the wire, breaking the connection.

MOTOR HUMS BUT WON'T RUN

Either the capacitor is bad or the impeller is jammed. Spin the shaft. If it won't turn freely, open the pump and clear the obstruction. If it does spin, check the capacitor.

The best way to check a capacitor is to replace it with a new one. Check the capacitor (see Fig. 4-7) for white residue or liquid discharge. Either is a symptom of a bad capacitor. There is a screw or clamp bracket that holds it in place and two wires, connected to the capacitor with bayonet-type clips. When you see it, you'll realize that not much instruction is needed.

All of this assumes your internal motor switch connection is set for 120 volts, for incoming power of 120 volts, or set for 220 volts if the incoming is 220 volts. Check the wiring diagram and power supply.

A more rare condition that might cause the motor to hum but not run is that your line voltage is not what it should be. Your 120 volts, for example, might be coming in at only 100 volts because of a faulty breaker or a supply problem from your power company. Use your multimeter to test the actual voltage supply at the motor.

THE BREAKER TRIPS

Disconnect the motor and reset the breaker. Turn the motor switch (or time clock on switch) back on and if it trips again, the problem is either a bad breaker or, more likely, bad wiring between the breaker and motor. Be very careful with this test. Switching the power back on with no appliance connected means you are now dealing with bare, live wires. Be sure no one is touching them and that they are not touching the water, each other, or anything else.

If the breaker does not trip when conducting this little experiment, the motor is bad. This usually means there is a dead short in the windings and the motor needs to be replaced. Water can cause this.

At Charles Bronson's pool, I found a large lizard had crawled into the motor housing through the air vents and when the timer turned the system on, the poor reptile became the short across the winding wires, burning out the motor.

LOUD NOISES OR VIBRATIONS

This is most often caused by worn out bearings. Take the pump apart and remove the load (impeller and water). If the motor still runs loud or vibrates, it is the bearings. Take it to a motor shop, or better still, replace the motor (unless the motor is relatively new or is still under warranty). This problem can also be caused by a bent shaft, although that is not common.

Not all noise is caused by the motor. Track down noises by a process of elimination, experimenting with various pieces of equipment (such as automatic pool cleaners, automated valves, heaters) all turned off, then turned on one at a time. See the sidebar on p. 124 for a list that will help find other culprits.

Priming the Pump

Sometimes the most difficult step is getting water moving through the pump. Priming means getting water started, creating a vacuum so more will follow.

TRICKS OF THE TRADE: NOISE CHECKLIST

Security

- **Is the pump properly secured to the deck or mounting block and is the mounting block secure?**
- **Are check valves rattling?**
- **Are pipes loose and vibrating? Grab sections of exposed pipe and see if the noise changes.**

Cavitation

- **Are suction and return line valves fully open or open too much?**
- **Is the suction-type automatic pool cleaner starving the pump?**
- **Undersized suction plumbing?**

Air

- **Is the pump strainer basket clean and the lid tightly fastened?**
- **Is the skimmer clogged or the water level low?**

Other troublemakers

- **Is the equipment located in a sound-magnifying environment, such as large concrete pad and masonry walls? Consider a vented "doghouse" cover.**
- **Is the heater "whining"? (See heater chapter.)**
- **Are loose filter grids rattling inside the filter canister?**

BASIC PRIMING

RATING: EASY

Let's go through the steps to prime most pools and spas.

1. **Water Level** Before starting a pump that you have had apart, always make sure there is enough water in the pool to supply the pump. In taking equipment apart, water is usually lost in the process and there might not be enough to fill the skimmer. I have also encountered pools that seem to have enough water, but will not prime unless filled to the very top of the skimmer. Sometimes that extra inch or two (2.5 to 5 centimeters) is enough to change the hydraulics of the system and get

it working. Factors such as distance of equipment from the pool and height above the pool also enter into the equation.

2. **Check the Water's Path** Often, priming problems are not related to the pump, but to some obstruction. Check the skimmer throat for leaves, debris, or other obstructions. Next, open the strainer pot lid, remove the basket, and make sure there are no obstructions or clogs in the impeller. Last, make sure that once the pump is primed it has somewhere to deliver the water. In other words, be sure all valves are open and that there are no other restrictions in the plumbing or equipment after the pump. If all this checks out, proceed.

3. **Fill the Pump** Always fill the strainer pot with water and replace the lid tightly so air cannot leak in. Keep adding water until the pot overflows so you fill the pipe as well as the pot. Sometimes the pump is installed above the pool water level so you will never fill the pipe (unless a check valve is in the line as well). Just fill what you can and close the lid.

4. **Start Up** Start the motor and open the air relief valve on top of the filter. Give the pump up to two minutes to catch. Carroll O'Connor's pool in Malibu took 4 minutes and 30 seconds to catch prime—you could set your watch by it (this is because it was about 4 feet higher than the pool level). Most pumps will catch prime sooner, and you don't want to overheat a dry-running pump.

Sometimes repeating this procedure two or three times will get the prime going. If there is a check valve in a long run of pipe, each successive filling of the pot pulls more and more water from the pool, which is held there each time by the check valve. Also, if it is warm outside, the air in the pipe might expand and create an airlock. The repeated procedures might finally dislodge the air.

THE BLOW BAG METHOD

RATING: ADVANCED

When basic priming fails, try a drain flush bag, also called a blow bag (Fig. 4-11). The drain flush is a canvas or rubber tube that screws onto the end of your garden hose. Slip this into the skimmer hole that feeds the pump and turn on the hose.

FIGURE 4-11 Using a drain flush bag.

The water pressure makes the bag expand and seal the skimmer hole so the water from the hose cannot escape and must feed the pump. After running the hose a minute or two, turn on the pump. When air and water are visible returning to the pool, pull the drain flush bag out quickly, while the pump is running, so pool water will promptly replace the hose water.

FILTER FILLING METHOD

RATING: PRO

Another method is what I call *filter filling*. Open the strainer pot, turn on the motor, and feed the pot with a garden hose. Open the filter air relief valve and keep this going until the filter can is full (water will spit out of the air relief valve).

Close the air relief valve, turn off the motor and garden hose, and quickly close the strainer pot. Open the air relief valve. The filter water will flood back into the pump and the pipe that feeds the pump from the pool. When you think these are full of water, turn the motor back on. The pump should now prime.

DETECTING AIR LEAKS

RATING: EASY

Okay, so none of this works. The problem might be that the pump is sucking air from somewhere, meaning it will not suck water (which is harder to suck).

Air leaks are usually in strainer pot lid O-rings, or the pot or lid itself has small cracks. The gasket between the pot and the volute might be dried out and leaking. Of course, plumbing leading into the pump might be cracked and leaking air.

If any of these components leak air in, they will also leak water out. When the area around the pump is dry, carefully fill the strainer pot with water and look for leaks out of the pot, volute, fittings, and pipes. Another way is to fill and close the pot, then listen for the sizzling sound of air being sucked in through a crack as the water drains back to the pool.

Sometimes there's just no easy answer—remove the pump and carefully inspect all the components, replace the gaskets and O-rings, and try again.

T-HANDLES

RATING: EASY

Many pumps employ threaded T-shaped bolts that secure the lid to the strainer pot (Fig. 4-3). Sometimes these corrode and snap off, with part of the bolt in the pot and the other part in your fist.

If part of the broken bolt extends above or below the female part on the pot, try using pliers, especially Vise-Grips, to grasp the broken section and twist it out. You can buy new T-bolts at the supply house. You can also take a flat-blade screwdriver and place it on the broken bolt end, tap it with a hammer to create a slot in the end of the broken piece, and twist out the broken piece like removing a screw.

If this doesn't work, take your tap and die set or electric drill and tap a small hole inside the broken piece, then use your Phillips-head screwdriver to grip inside the hole and twist out the broken piece. If all

---

FAQS: PUMPS AND MOTORS

Will my pump be damaged if it runs dry?

- After several minutes of running dry, seals and plastic components will begin to warp, resulting in leaks later. Never run a pump dry for more than a minute.

Should the pump be running when swimmers are in the pool?

- Bather load will quickly turn the water cloudy, so circulation is a key to a clean pool. However, be careful if you have a main drain and/or small children. Be sure all bathers know that suction at the skimmer and main drain can cause injury if hands, feet, or hair are allowed to block them off while the pump is running.

How much does it cost to run the pump?

- Depending on the cost of electricity in your area and the number of hours you need to run the pump to keep your pool clean, it can cost from pennies a day up to $2 a day. It will rarely exceed the higher figure, even in summer.

Is there such a thing as a silent pump/motor?

- A properly functioning motor that is not laboring against blockages, such as a dirty filter or clogged skimmer, should not be objectionably loud. Loud motors are a sign of worn bearings or obstructions in the circulation system. You can also cover your motor with specially designed housings that reduce noise.

---

else fails, remove the pot from the pump and take it to a machine shop to be tapped out and rethreaded.

A hint: If one of these handles breaks off on Friday afternoon, you can clamp that side of the lid on the pot with your Vise-Grips and run the system until Monday when you can get the parts to fix it.

Cost of Operation

As with any electrical appliance, you can easily calculate the cost of operating a pump. Electricity is sold by the kilowatt-hour. This is 1000 watts of energy each hour. You know that volts × amps = watts, so you can look at the motor nameplate and see that the motor runs, for example, at 15 amps when supplied with 110-volt service, and 7 amps when supplied by 220-volt service.

Let's say the pump in our example is running on 220-volt service—220 volts × 7 amps = 1540 watts. Looking at an electric bill, you learn that you pay 15 cents per kilowatt-hour. As noted, a kilowatt is 1000 watts, so if you divide 1540 watts by 1000, you get 1.54. That is multiplied by your kilowatt rate (15 cents), equaling 23 cents for every hour you run the appliance. If you run the motor 8 hours per day, that means 23 cents × 8 hours = $1.84/day. Over a month, that equals 30 × $1.84 = $55.20/month.

Filters

As the name implies, the filter is the piece of pool equipment that strains impurities out of water that is pumped through it. There are few moving parts (in fact no parts should be moving when the filter is in operation) and simple components.

Types

Three basic types of filter are in common use and each is preferred for various reasons. Each type is more efficient for various adaptations and you might also discover regional preferences around the world. For example, in regions where sand is plentiful but diatomaceous earth is scarce, the sand or cartridge filter will be more commonly in use than the DE filter. Setting aside such regional prejudice, however, as I review the types of filters, I will also discuss why each is preferred based on technical application.

Diatomaceous Earth Filters

In the diatomaceous earth type of filter, also called a *DE filter* (Fig. 5-1), the water passes into a metal or plastic tank, through a series of grids (also called *filter elements*) covered with fabric, and back out of the unit. The grids do not actually perform the filtration process, but instead are coated with filter *media*, diatomaceous earth, that does the actual filtering work.

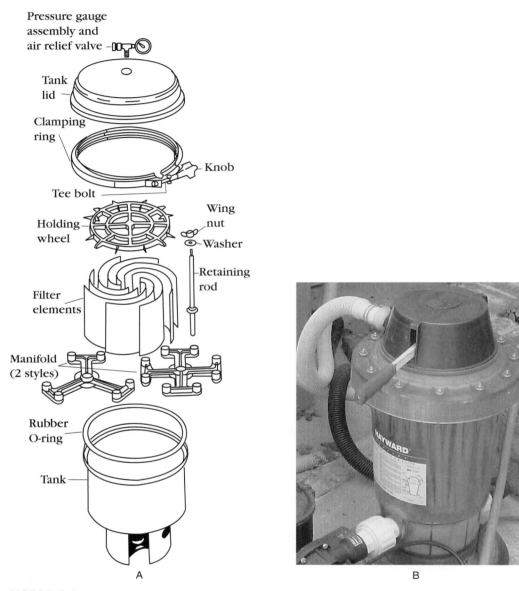

Pressure gauge assembly and air relief valve

Tank lid

Clamping ring

Knob

Tee bolt

Holding wheel

Wing nut

Washer

Retaining rod

Filter elements

Manifold (2 styles)

Rubber O-ring

Tank

A

B

FIGURE 5-1 (A) Diatomaceous earth (DE) filter; (B) DE filter with tubular grids. *A: Pentair Pool Products, Inc.*

DE is a white powdery substance found in the ground in large deposits. It is actually the skeletons of billions of microscopic organisms that were present on earth millions of years ago. So, in essence, this powder is akin to dinosaur bones. If you look at DE under a microscope, you will see what appear to be tiny sponges, thus the filtration ability becomes more apparent. Just like a sponge, water can pass through, but the impurities in the water can't. Because the DE particles are so fine, they can strain very small, actually microscopic, particles from the water as it passes through.

So why not just dump a few pounds of DE in a tank and pass the water through? Good question. Because the DE will collapse together and cake, making it impossible for even the water to get through. Therefore, the tank is equipped with free-standing grids (also called *elements*) that are coated with DE to accomplish the filtration process. DE filters are usually located on the return side of the pump (a pressure filter).

As can be seen in Fig. 5-1A, the grids (elements) are mounted on a manifold and the resulting assembly fits into the tank. A retaining rod through the center screws into the base of the tank and a holding wheel keeps the grids firmly in place. The top of the tank is held in place by a clamping ring, and the two parts are sealed with a thick O-ring to prevent any leaking.

The water enters the tank at the bottom and flows up around the outside of the grid assembly. It must flow through the grids, down the stem of each grid, and into the hollow manifold, after which it is sent back out of the filter. This type of DE filter is called a *vertical grid*, for obvious reasons.

One way to clean this filter is called *backwashing*, a concept I will discuss in more detail later. As the term suggests, backwashing means the water is redirected through the filter in the opposite direction from normal filtration (accomplished with a backwash valve, also discussed later), thereby flushing old DE and dirt out of the filter.

Not all vertical DE filters are equipped with a backwash valve to allow backwash cleaning. These types must be disassembled each time for cleaning. Some are also equipped for *bumping* rather than backwashing. In this rather ridiculous process, the dirty DE is bumped off the grids, mixed inside the tank so the dirt is evenly distributed within the DE material, then recoated onto the grids.

Figure 5-1B shows a DE filter that employs the bumping method, but instead of traditional rigid grids, it uses flexible tubes made of a foamlike material. The tube grid design packs more filtration surface area into a filter than the rigid grid counterpart, so it's a nice concept for above-ground pool equipment packages, for which space is often an issue.

The idea is that since most dirt sits on the outermost layer of DE as the DE rests on the grids, there is still relatively clean, unused DE available. By mixing DE and dirt together, this type of filter design reasons that somewhat cleaner DE can be brought to the surface to prevent the need to actually replace it. Because I find this type of filter to be too stupid for words, it will not be discussed further.

In some areas, out of concern that DE will clog pipes, local codes require that DE not be dumped into the sewer system. In this case, a *separation tank* is added next to the filter (Fig. 5-2). When draining or backwashing the filter, the dirty water is passed through a canvas strainer bag inside this small tank before going into sewer or storm drains. The canvas bag strains most of the DE out of the water so it can be disposed of elsewhere.

Some rather tortuous vertical DE filter units were designed with the manifold on top and the holding wheel on the bottom. My experience is that these do not work any better but are **much** harder to get back together after disassembly for cleaning.

All filters are sized by the square footage of surface area of their filter media. In DE filters therefore, the total surface square footage of the grids would be the size of the filter. Typically there are eight grids in a filter totaling 24 to 72 square feet (2 to 7 square meters), designed into tanks that are 2 to 5 feet (60 to 150 centimeters) high by about 2 feet (60 centimeters) in diameter. Obviously the larger the filter, the greater capacity the filter will have to move water through it. Therefore, filters are also rated by how many gallons per minute can flow through them. More on this later.

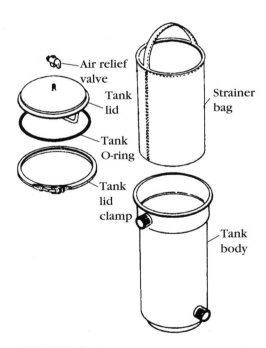

Air relief valve

Tank lid

Strainer bag

Tank O-ring

Tank lid clamp

Tank body

FIGURE 5-2 **Separation tank.** *Premier Spring Water, Inc.*

Sand Filters

If the filter media is not DE, it must be sand. Sand and gravel are the natural methods of filtering water.

The concept of the pressure sand and gravel filter is simple. Water is passed through a layer of sand and gravel inside a tank, which strains impurities from the water before it leaves the tank. I designate these types of filters as *pressure* sand and gravel because like their DE cousins, the water is under pressure inside the tank from the resistance created by trying to push it through the filter media. This differentiates it from another type of sand and gravel filter used especially in fish ponds where there is no such pressure. Pressure sand filters are also called *high-rate sand filters*.

In Fig. 5-3A, the water enters the tank through the valve (item 7) on top and sprays over the sand inside. The water then runs through the sand, impurities being caught by the sharp edges of the grains, and is pushed through the manifold at the bottom where it is directed up through the pipe in the center and out of the filter through another port of the valve on top. The individual fingers of the drain manifold (item 20) are called *laterals,* and the center pipe used to return clean water (item 19) is called a *stanchion pipe*. To drain the tank, a drain pipe is provided at the bottom as well. The design shown in Fig. 5-3B works in a similar manner, but the water enters from the side and flows to the top of the filter to be sprayed over the top of the sand. The water percolates through the sand and enters the laterals before moving up the center pipe and out the side. In this filter, the flow control valve is side mounted.

Sand filters are also sized by square footage and gallons per minute. Does some engineer measure the surface area of the billions and billions of grains of sand in a given filter to arrive at the total square footage of filtration area? No, actually it is based on the surface area of the sand bed created inside the filter. Since most sand filters are round, a filter measuring 24 inches (60 centimeters) in diameter with a radius of 12 inches (30 centimeters) would have a sand bed of 3.1 square feet (0.29 square meter) (using the formula pi × radius squared). Knowing the volume of sand recommended for any given filter (expressed in cubic feet or cubic meters), the manufacturer arrives at a square footage value and a resulting gallons per minute (or liters per minute)

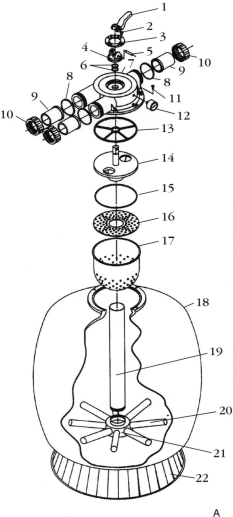

1 Backwash valve handle
2 #8 retaining screw
3 Indicator plate
4 Handle bracket
5 Handle pin
6 Rotor shaft O-ring
7 Valve body with gasket
8 Plumbing adapter O-ring
9 Plumbing adapter
10 Plumbing adapter nut
11 $1/4$" assembly bolt with washer
12 0–60 psi pressure gauge
13 Backwash valve/tank O-ring
14 Rotor plate
15 Rotor plate O-ring
16 Diffuser
17 Distributor/strainer
18 Filter tank body
19 Center water flow pipe
20 Lateral underdrain (8)
21 Underdrain manifold hub
22 Filter tank stand

A

FIGURE 5-3 **High-rate sand filters: (A) top valve; (B) side valve.** *Pentair Pool Products, Inc.*

rating. Sand filter tanks are usually large round balls, anywhere from 2 to 4 feet (60 to 120 centimeters) in diameter.

Cartridge Filters

A cartridge filter is similar to a DE filter except there is no DE filter media. In Fig. 5-4, water flows into a tank that houses one or more

1   Air relief valve
2   Adapter
3   Pressure gauge
4   O-ring
5   Nut
6   Lid
7   O-ring
8   Strainer
9   Air relief tube
10 Diffuser assembly
11 Upper pipe assembly
12 Air relief tube connector
13 Lower pipe assembly
14 Tank and foot assembly
15 Laterals
16 Lateral hub
17 Drain spigot
18 Drain plug
19 Lid wrench
20 Lock nut
21 Spacer
22 O-ring
23 Spacer
24 Gasket
25 Bulkhead
26 O-ring
27 Washer
28 Base
29 O-ring
30 2" threaded adapter kit
31 1.5" threaded adapter kit
32 (See 30, 31)
33 Closure kit
34 Fitting package
35 Spacer
36 Bulkhead kit

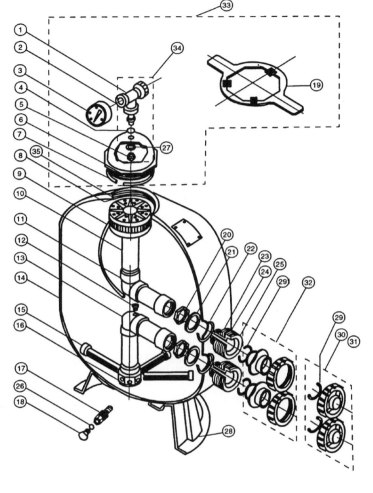

B

FIGURE 5-3 *(Continued)*

cylindrical cartridges of fine-mesh, pleated fabric (usually polyester). The extremely tight mesh of this fabric strains impurities out of the water.

Figure 5-5 also shows a cartridge filter, but instead of one large cartridge, this model uses several smaller ones. Figure 5-4B shows a cartridge filter mounted right in the skimmer. Cartridge filters are classified by

1 Pressure gauge
2 Filter lid assembly
3 Tank O-ring
4 Tank body
5 Water diverter
6 Drain cap with
   gasket
7 Safety latch
8 Filter cartridge
9 Air bleed tube
10 Air release valve
   with O-ring
11 Locking ring

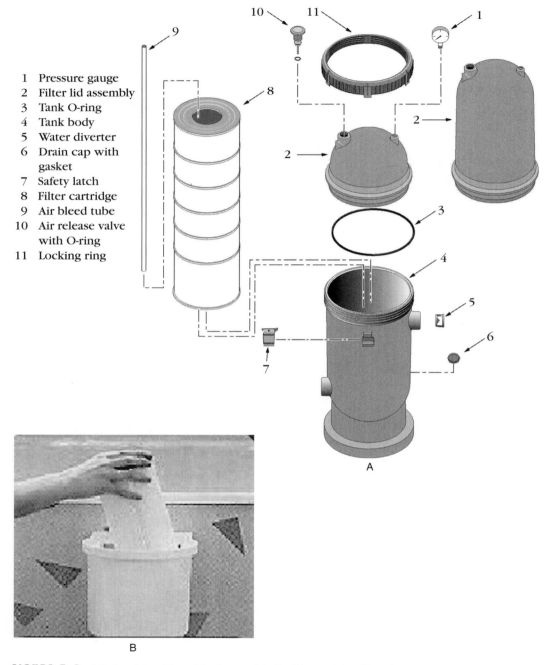

FIGURE 5-4 **(A) Cartridge filter (single cartridge); (B) cartridge filter in skimmer.** *A: Sta-Rite Industries. B: Zodiac American Pools, Inc.*

square footage of filter surface as are DE and sand filters. By pleating the cartridge material, a lot of square footage can fit into a very small package. This would not be possible if DE were added, because the tight pleating would prevent DE from coating evenly and effectively. Typically, cartridge filters for residential use range from 20 to 120 square feet (2 to 12 square meters) in a tank not more than 4 feet (1.2 meters) high by 3 feet (1 meter) in diameter.

Makes and Models

Now that you have had a brief introduction to typical pool filter types, how do you make a selection and what size do you need for a given pool?

FIGURE 5-5 Cartridge filter (multiple cartridges).

Sizing and Selection

A good turnover rate (if you've forgotten turnover rate, see that section) requires a pool to be turned over in 6 hours, and a wading pool in 1 hour. Using this information and knowing the total gallons in your pool, you will know how many gallons per minute the filter must be able to handle.

As an example, my pool is 40,000 gallons. To turn that over every 6 hours, I need a filter that can handle

$$40,000 \text{ gal } (151,400 \text{ L}) \text{ in } 6 \text{ h} = 40,000 \text{ gal in } 360 \text{ min}$$
$$= 111.11 \text{ gal } (420 \text{ L}) \text{ per minute}$$
$$(40,000 \div 360)$$

So with this information, you compare the required gallons per minute for our sample pool, 111, to the flow rate of the filter you plan to buy.

Another sizing and selection criteria is dirt. How dirty does the body of water get that this filter must service? You want to oversize the filter a bit so you don't have to clean it out every day, week, or hour (choose one). This time between filter cleanings is called the *filter run*.

So you might as well get a filter 20, 50, or 100 percent larger than you actually need so it can hold more dirt and thereby leave more time between cleanings—right?

Well, not exactly. If the pump cannot deliver the gallons per minute for the square footage of this huge filter you have just installed, only part of the tank will fill with water, effectively giving you a smaller filter anyway. Also, the pump won't backwash the filter completely if it can't match the gallons per minute rating. So the obvious question is, which do we use? You might also want to consider price and efficiency.

Price is a factor of the current market and supplier availability in your area, so I'll let you check that one out for yourself. If you just want the filter that does the best job regardless of price, then the answer lies in the mighty micron. A *micron* is a unit of measurement equal to one millionth of a meter or 0.0000394 inch. Put another way, the human eye can detect objects as small as 35 microns; a talcum powder granule is about 8 microns, and a grain of table salt is about 100 microns.

So what? Well, each filter type has the ability to filter particles down to a particular size, as measured in (you guessed it) microns. Various references and manufacturers will give you different estimates and they might all be right as you will see in a minute. However, here's a rule of thumb:

- Sand filters strain particles down to about 60 microns
- Cartridge filters strain particles down to about 20 microns
- DE filters strain particles down to about 7 microns

I read an article recently (with data supplied by a manufacturer, of course) claiming sand filters at 25 microns, cartridges at 5 microns, and DE at 1 micron. Gentlemen, modesty, please!

Yet, as I said, both estimates are correct. Sand filter efficiency is based on the age of the sand. New sand is crystalline and many-faceted. These sharp-sided facets are what actually catch debris out of the water and filter it. As sand ages, the years of water passing over it cause erosion of those facets until each grain of sand becomes smooth and rounded, losing its ability to trap particles of the smallest size. So a new sand filter with fresh sand might strain particles of 25 microns, while a unit in service one year might be lucky to strain particles of 60 microns.

Efficiency of a DE filter is affected by the cleanliness of the DE. As DE becomes dirtier, the filter will actually trap finer particles, but the flow rate will decrease, thus making it less efficient overall.

Because cartridges do not rely on organic material like sand or DE, they are not affected the same way. As long as the cartridge surface area remains clean, it will filter the same size particles (although extremely old cartridges that have been acid washed many times will stretch out

> ## TRICKS OF THE TRADE: SAND FILTERS
>
> - **To improve sand's ability to filter, add ½ cup or 125 milliliters of alum (aluminum sulfate) for every 3 square feet or 0.3 square meter of the filter's rated size. This acts like a DE coating on the grids of a DE filter and helps the sand strain finer particles.**
> - **Choose a sand filter for pools with heavy use. Extreme amounts of dirt and waste will quickly clog a cartridge or DE filter and take more time to clean than the simple backwashing procedure for sand filters.**

somewhat, creating a mesh that is not so fine as when new). However, dirt does affect the cartridge filter, as it affects all filters, but with the cartridge it actually has a positive effect.

Dirt makes the mesh of the polyester cartridge material even tighter, essentially acting like DE. I know some service technicians who add a small handful of DE to a cartridge filter after cleaning it to start this process. Thus, with each successive pass of water through the cartridge filter, it becomes more efficient as the mesh gets finer and finer from the retained particles. This is also true with sand filters, up to a point.

Regardless of these variations, the one consistent fact you will note in the estimates is that sand filters are the least efficient filters in terms of how fine the particles are that they can strain from the water; cartridge filters appear to be the second best and DE the best. I say *appear to be* because a slightly dirty cartridge filter will strain particles of 5 to 10 microns, while a dirty DE filter simply clogs up and doesn't allow as much water to pass through as it should. Although it too is straining particles of 5 to 10 microns in size, if the flow is reduced because the DE is clogged, what good is the rated straining capability?

This is why cartridge filters are generally regarded as the best overall filtration method. They are easy to clean and require no added organic media.

Backwash Valves

As I noted previously, backwashing is a method of cleaning a DE or sand filter (cartridge filters do not backwash) by running water backwards through the filter, flushing the dirt out to a waste line or sewer line. There are basically two types of backwash valves—the piston and the rotary (the multiport being a variation of the rotary valve).

PISTON VALVE

Figure 5-6 shows a piston-type backwash valve. In the normal operating position, the water enters the valve (marked Pump discharge)

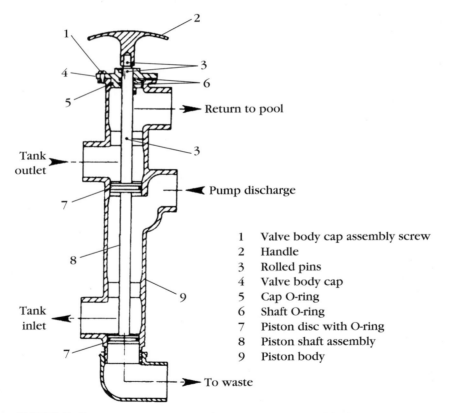

1	Valve body cap assembly screw
2	Handle
3	Rolled pins
4	Valve body cap
5	Cap O-ring
6	Shaft O-ring
7	Piston disc with O-ring
8	Piston shaft assembly
9	Piston body

FIGURE 5-6 Piston backwash valve (normal operating position).

and, as you can see, the piston discs (item 7) only allow the water to go out to the filter tank inlet. The water is filtered through the sand or DE and returns at the top of the valve (marked Tank outlet). Again, you can see that the piston disc (item 7) only allows the water to flow to the outlet of the valve (marked Return to pool). Normal filtration has occurred.

To backwash the filter and flush out the dirt, the handle of the piston assembly (item 2) is raised. Figure 5-7 shows the same valve with the piston raised to the backwash position. Figure 5-6, item 3, shows two rolled pins that ensure exact placement of the piston inside the body. The outside pin acts as a stop when pushing the valve down, the inside pin acts as a stop when pulling it up.

Notice that the water still enters from the pump, but now the piston discs force the water to flow *in* to the filter tank through the *outlet* opening. The water flows backward through the filter, flushes the dirt out of the tank and *out* of the valve *inlet* opening. As you see, once inside the valve again, the dirty water is directed to the line marked Waste. O-rings on the piston discs ensure that water does not bypass the discs.

By the way, never change the piston position when the pump is running, because this puts excessive strain on the pump and motor and the O-rings of the valve, causing it to leak. The piston-type backwash valve is usually plumbed onto the side of the filter tank.

ROTARY VALVE

The rotary backwash valve, used only on vertical DE filters, does the same thing as the piston valve and in pretty much the same way. Figure 5-8 shows the typical rotary valve. Changing the direction of water flow is done by rotating the interior rotor (item 11 for the bronze, or

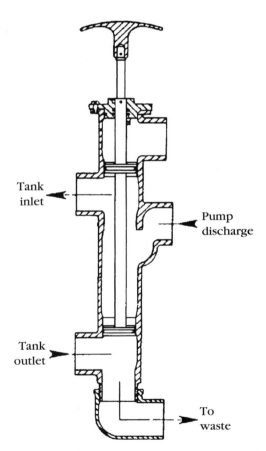

FIGURE 5-7 Piston backwash valve (backwash position).

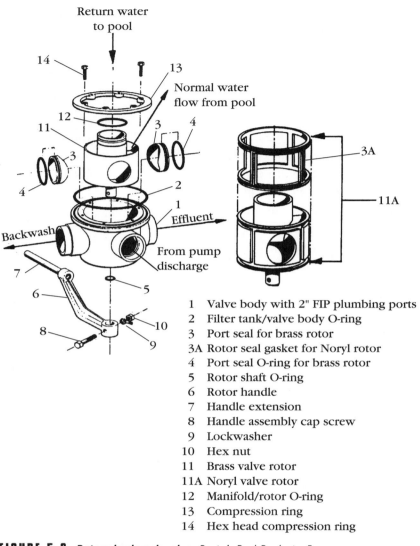

Return water
to pool

14

13

Normal water
flow from pool

12

11

3   4

3

4

2

1
Effluent

Backwash

From pump
discharge

7

6

5

8

10

9

3A

11A

1   Valve body with 2" FIP plumbing ports
2   Filter tank/valve body O-ring
3   Port seal for brass rotor
3A   Rotor seal gasket for Noryl rotor
4   Port seal O-ring for brass rotor
5   Rotor shaft O-ring
6   Rotor handle
7   Handle extension
8   Handle assembly cap screw
9   Lockwasher
10   Hex nut
11   Brass valve rotor
11A Noryl valve rotor
12   Manifold/rotor O-ring
13   Compression ring
14   Hex head compression ring

FIGURE 5-8 Rotary backwash valve. *Pentair Pool Products, Inc.*

item 11A for the plastic version). A rotor gasket seal (item 3 or 3A) or
O-rings (item 4) keep water from leaking into the wrong chamber.

This type of valve is mounted underneath the filter tank. The tank
has a hole in the bottom to accommodate the unit. The valve body
(item 1) is held under the tank while a compression retaining ring
(item 13) is placed inside the tank and the two halves are bolted

together with cap screws (item 14). The handle underneath allows you to rotate the interior rotor to align it with the openings of the valve body as desired.

In normal filtration, water comes into the valve through the opening marked From Pump Discharge and flows up into the filter through the large opening on top of the rotor marked Normal Water Flow. After passing through the DE-coated grids, the water flows back through the inside of the grids, into the manifold (see the DE filter section), down through the center of the rotor (marked Return Water), then out the effluent opening and on to the heater or back to the pool.

When the valve is rotated, water is sent up through the middle of the rotor, up inside the grids. The dirt and DE are pushed off the grids and the water flows from inside the grids to the outside. This is flushed back through the rotor, opposite the normal flow, and directed to the opening marked Backwash. This line is connected to a waste or sewer line. As with all backwash valves, do not operate it while the pump is running or you might damage the valve and the pump and motor.

MULTIPORT VALVE

The multiport backwash valve is used on sand filters and looks just like a rotary valve when disassembled, except that there is one more choice for water flow. A rinse is added, so that after the water has backwashed through the filter, clean water from the pump can be directed to clean out the pipes before returning to normal filtration. This prevents dirt in the lines from going back to the pool after backwashing. If you understand piston and rotary valves, you will have no trouble with a multiport valve when you encounter it in the field.

Backwash Hoses

To accomplish the noble purpose of backwashing, the dirty water has to go somewhere. Some waste or backwash openings are plumbed directly into a pipe that sends the water into a nearby deck drain and on toward the sewer. Many are not hard-plumbed and you must connect a hose to direct the waste water wherever you want it to go.

By the way, dirty filter water is an excellent fertilizer for lawns because it is usually rich in biological nutrients, algae, decaying matter, and DE (which, as you now know, is a natural material). You can run your backwash hose on the lawn or garden, providing the water chlorine residual level is not above 3 ppm (see Chap. 8). Chlorine levels higher than that might burn the grass.

A backwash hose can be your pool vacuum hose, clamped onto the waste line of the backwash valve with a hose clamp. Normally, however, a cheap, blue, collapsible plastic hose is attached to the waste opening with a hose clamp. This 1½- or 2-inch-diameter (40- or 50-millimeter) hose is intentionally made flimsy because the water is not under much pressure when draining to waste and also to allow it to be easily rolled up and stored near the filter. This ease of rolling up is an advantage because backwash hoses normally come in lengths of 20 to 200 feet (6 to 60 meters)—you might have to route waste water into a street storm drain, so very long lengths are common.

Pressure Gauges and Air Relief Valves

Most filters are fitted with a pressure gauge, mounted on top of the filter (Fig. 5-4, item 1). Sometimes the gauge is mounted on the multiport valve (Fig. 5-3, item 12). These gauges read 0 to 60 psi (0 to 4000 millibars) and are useful in several ways.

TRICKS OF THE TRADE: THE POOL BLOOD PRESSURE MONITOR

- When a filter is newly put into service or has just been cleaned, I make it a habit to note the normal operating pressure (in fact, I carry a waterproof felt marking pen and write that pressure on the top of the filter). Most manufacturers tell you that when the pressure goes more than 10 pounds (700 millibars) over this normal operating pressure it is time to clean the filter.

- The other value of the pressure gauge is to quickly spot operating problems in the system. If the pressure is much lower than normal, something is obstructing the water coming into the filter (if it can't get enough water, it can't build up normal pressure). If the gauge reads unusually high, either the filter is dirty or there is some obstruction in the flow of water after the filter.

- When the pressure fluctuates while the pump is operating, the pool or spa might be low on water or have some obstruction at the skimmer—when the water flows in, the pressure builds, then as the pump sucks the skimmer dry, the pressure drops off again. This cycling will repeat or the pressure will simply drop altogether, indicating the pump has finally lost prime.

Mounted on a yoke or T fitting along with the pressure gauge, you will normally find an air relief valve (Fig. 5-4, item 10). It is simply a threaded plug that when loosened allows air to escape from the filter until water has fully filled the tank. When a system first starts up, particularly after cleaning or if the pump has lost prime, there is a lot of air present in the filter. If it is left like that, the filter might be operating at only half its capacity.

Figure 5-5 shows an inside view of a cartridge filter. The area above the cartridge is called the *freeboard*. This empty area is present above the filter media of all filters. For some of this to be air rather than water is okay, but if the air in the tank were down to, for example, the halfway mark of the cartridge and water were flowing around only the lower half of the cartridge, you would effectively cut the filter square footage in half.

> **TOOLS OF THE TRADE: FILTERS**
> - **Heavy flat-blade screwdriver**
> - **Hacksaw**
> - **PVC glue**
> - **PVC primer**
> - **Pipe wrench**
> - **Socket wrench set**
> - **Teflon tape**
> - **Silicone lube**
> - **Needle-nose pliers**
> - **Hammer**
> - **Emery cloth or fine sandpaper**
> - **Garden hose**
> - **Vise-Grips**

So it is important to let the air out of the filter at every service call. Again, some air is normal, so don't be obsessive about it; just be sure that the filter media itself is covered with water, not air. Newer filters now have automatic air relief valves, leaving one less thing to chance or checklist.

Repair and Maintenance

As noted previously, there are no moving parts on a filter when it operates and few when it is at rest. Therefore, there's not much to break down and, when they do, filters are easy to repair.

Installation

RATING: EASY

Replacing a filter is one of the easiest repair jobs you can get. There is no electricity or gas to hook up and normally you are dealing with only three pipes.

TRICKS OF THE TRADE: FILTER INSTALLATION

- On a filter with threaded openings in a plastic base (like most cartridge filters) or a plastic multiport valve (like most sand filters), be careful not to overtighten—you'll crack the plastic. On a vertical grid DE filter with the rotary valve on the bottom, lay the filter down and screw the MIPs into place. Some manufacturers make this a knuckle-busting experience. Be sure you have the fittings tight—if they leak, you'll have to cut out all your work to fix it.

- Some manufacturers still provide filters with bronze plumbing openings on the backwash valve for sweating copper pipe. If you encounter this type of plumbing, remove any parts that might suffer from heat when you sweat the fittings in place. Plastic grids and manifolds will melt if they are not removed, and with the lid on the filter, the heat inside will build up quickly. Just as the backwash valve conducts water efficiently, so too will it conduct heat to those plastic parts.

- If a check valve was not a part of the plumbing between the filter and the heater, it is not a bad idea to add one now. If there is any chance of hot water flowing back into the filter, you run the risk of melting the grid or cartridge materials. Also, you don't want the heater to sit without water in the heat exchanger (see the heater chapter).

1. **Shutdown** Turn off the pump and switch off the circuit breaker. This way you can make sure it won't come back on (from the time clock, for example) until you're ready.

2. **Drain** Drain the old filter tank by opening the drain plug or backwash valve. If you don't do this first, you'll take a bath when you cut the plumbing.

3. **Cut Out** Cut the pipe between the pump and filter in a location that makes connecting the new plumbing easiest. There's no rule of thumb here, just common sense. If the original installation has more bends and turns in the pipe than needed, now is a good time to cut all that out and start over. Eliminating unnecessary elbows increases flow and reduces system pressure. Cut the pipe between the old filter and the heater using the same guidelines. Last, cut the waste pipe if there is one plumbed into a drain.

4. **Remove** Remove the old filter. Even without water, the old filter will be heavy, so you might want to disassemble it for removal. Sand filters are the worst, and you will have to scoop out the

heavy, wet sand before you will budge the tank. Save any useful parts. Old but still working grids, valves, gauges, air relief valves, lids, lid O-rings, cartridges, and other components make great emergency spares.

5. **Installation** Set up the new filter. After removing it from the box, make sure all the pieces are there—grids, pressure gauge, etc. Most new filters come with instructions, and it really pays to read these. While the unit is out in the open and easy to work on, screw into place the appropriately sized mip fittings [1½ or 2 inch (40 or 50 millimeter)], after applying a liberal coating of Teflon tape or pipe dope (see the plumbing section). Some manufacturers include their preference of tape or pipe dope right in the box.

6. **Plan the Plumbing** Place the new filter in the location by the pump and figure out the plumbing between the pump and filter, between the filter and heater, and between the filter and waste line (if appropriate). Some creativity and planning here will save many service headaches later. Basically you want to avoid elbows and you want to leave enough room between the pump, filter, heater, and pipes for service access. Remember, you'll have to clean this filter someday. Can you easily access the lid? Will water flowing out of the tank as you clean it flood the pump? Can you access the backwash valve and outlet as needed?

7. **Plumb** Plumb it in. Use the plumbing instructions in the chapter on basic plumbing and make careful connections. Who was it that said, "There never seems to be enough time to do it right, but there's always enough time to do it over when it leaks"?

Starting up the newly installed filter is just like restarting after cleaning, so read further for start-up procedures and hints.

Filter Cleaning and Media Replacement

The most important thing you can do for a pool is to keep the filter clean. This is also the simplest way to ensure the other components work up to their specifications. Let's review the process for cleaning each type of filter.

QUICK START GUIDE: FILTER CLEANING (ALL TYPES)

1. **Prep**
   - Shut off pump and tape over switch or breaker so no one can turn it on before you finish.
   - Isolate equipment by closing valves at suction line (at skimmer and/or main drain connection before pump) and return line (at pool discharge outlet).

2. **Disassembly**
   - Remove filter top or lid.
   - DE filter: Carefully remove retainer, grids, manifold (Fig. 5-9A)
   - Cartridge filter: Remove cartridge.
   - Sand filter: Remove debris basket or other components to expose sand inside filter.

3. **Clean**
   - DE or cartridge: Hose off grids or cartridges thoroughly (Fig. 5-9B). Soak in cleaning solution if needed and rinse thoroughly. Rinse interior of filter and any other components.
   - Sand: Insert hose and flush dirt from sand, stirring it with a broom handle as you flush. Rinse until water flows out clean.

4. **Reassemble**
   - Rebuild components, tops, and lids in reverse order of disassembly procedure.

5. **Restart**
   - Reopen pool plumbing valves.
   - Start pump, purge air, and check for leaks.
   - DE only: Add DE at skimmer slowly, watching for any that passes back into pool (if DE enters pool after first 20 seconds of circulation, repeat steps 1 through 4).

DE FILTERS

RATING: EASY

I do not believe backwashing a DE filter is of any value as a regular practice and, in fact, I know it can be harmful to the filter and pool cleanliness. Obviously the makers of backwash valves and those who have bought into their technology over the years will disagree with me, but here's what I believe based on years of experience. When you backwash, some dirt and some DE are flushed from the filter. The remainder drops off the grids and falls to the bottom of the filter in clumps. The manufacturers say that after backwashing 70 percent of the DE has been removed, so you need to replace that amount. I have opened up filters that were just backwashed and have seen as little as 10 percent of the DE washed away, while others had virtually 100 percent washed away.

If you don't know how much went out, how do you know how much to put back in? If you add too little, the filter grids will quickly clog with dirt and the pressure will build right back up, even stopping the flow of water completely. If you add too much, you will get the same effect by jamming the tank with DE.

Moreover, backwashing cannot remove oils from the grids, which get there from body oil, oil in leaves, and suntan lotions. You can backwash for hours and when you open up the filter, you will find the grids clogged with oils and a layer of DE and dirt that sticks to these oils.

Finally, backwashing wastes water. If you break down the filter and clean it completely, you will use some water to wash the grids and tank, but you will not have to clean it again for weeks or even months.

When you backwash you really are not cleaning the filter thoroughly. You'll be backwashing again in a few days or weeks, and when you get tired of that you will break down and clean the filter anyway.

Okay, I've had my say! Now I'll tell you the one time when backwashing a DE filter is useful. You have a pool that has been trashed by winds, mudslides, algae, or other heavy debris. You start to vacuum it, and quickly the filter can't hold any more dirt. To save a lot of time, you backwash, add a little fresh DE, and get on with the job. You repeat this process until the big mess is cleaned up, then you break down the filter and clean it properly.

Another important fact about backwashing. Since the water is going inside the grid and flowing outward, any debris in the water from the pool will clog the inside of the grids (or laterals on sand filters), rendering them useless.

Cleaning: So how do you properly break down (or tear down) and clean a DE filter? I'm going to describe a common style of vertical grid tank DE filter (Fig. 5-1). They are common in the field and if you can do these, you can do them all.

TRICKS OF THE TRADE: FILTER CLEANING

- Some filters make such a tight seal with the O-ring and lid that a nuclear bomb will not remove them. Draining the tank first won't help—as the water drains out, it sucks the lid on even tighter. What I do is, after removing the clamping ring, turn on the pump for a few seconds. The pressure from the incoming water will pop the lid off. When it does, grab it quickly, otherwise when the pump is shut off and the water recedes, it will suck the lid right back onto the tank.

- Some manufacturers do not approve of this procedure. I have "popped" literally thousands of filter lids and never had one pop more than a few inches off the tank. I have also never observed damage to the equipment by this technique. The pressure is applied evenly as the lid pops off, so the tank or lid doesn't warp or bend. Be sure not to grasp the lid by the gauge assembly—they snap off easily.

- The only caveat to this practice is to be sure that you don't stand in the water that will inevitably flow out of the tank as the lid pops while you are holding onto the pump/motor switch. Water and electricity can be deadly. Also, if the motor is installed directly adjacent to the filter tank, the water flowing out might flood the motor. In that case, get a screwdriver and crowbar to remove the lid, or wrap the motor in a plastic bag.

1. **Shutdown** Turn off the pump and switch off the circuit breaker.

2. **Lid Removal** Remove the lid of the filter. That might sound easy, but depending on the design, it can be real work. On some filters, it is as easy as removing the clamping ring and applying light pressure under the lid with a screwdriver (be careful not to gouge the lid or O-ring—if you do, leaks will develop in these spots).

3. **Grid Removal** Open the tank drain and let the water run out. Remove the retainer's wing nut and remove the retainer (also called the holding wheel). Now gently remove the grids (elements). One design flaw of many grids lies in the fact that they are made like small aircraft wings—large, curving units—but they are set into the manifold on stubby little nipples. Applying a reasonable amount of force on the rather large wing part of the grid won't hurt it, but the resulting torque on the flimsy nipple will snap it right off. Therefore, to remove the grids, wiggle them gently from side to side as you pull them straight up and out (Fig. 5-9A).

Tennis star John McEnroe fired his pool man of many years and called my company. I went over and found the pool very dirty and the filter pressure almost off the scale. I popped the lid and found so much dirt and DE in the tank that there was literally no room for water. This condition is called *bridging* because the DE and dirt bridge the normal gaps between the grids, clogging and effectively reducing the amount of filter area. It was so packed that it took me three hours of archaeologist-like excavation with a small spade, stick, and lots of water to finally get the grids free. This, by the way, was the result of a pool guy who backwashed and added fresh DE about once a month—for 6 years. He never once opened the filter. The cumulative result was that he was adding more DE than was being flushed out, the pool stayed dirty because it couldn't filter or be vacuumed, and, best of all, he lost the customer.

The point of telling you this is that to some lesser degree you might find the same condition when you open a filter. Be prepared to hose out the tank while the grids are still in place (if the drain hole isn't also clogged with DE) or patiently excavate the dirt and DE until you can free the grids.

4. **Rod Removal** Remove the retaining rod. It threads into the base of the rotary valve like a screw, so just unscrew it. Sometimes it is corroded in place, so have pliers (Vise-Grips work best) handy to grip the rod and unscrew it. A word of caution—the rod

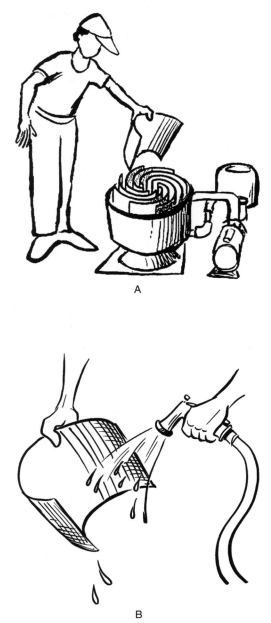

A

B

FIGURE 5-9 Filter cleaning: (A) step 1; (B) step 2.

might be corroded enough that if you force it with your pliers, it snaps off at the bottom, leaving the threaded end in the rotary valve.

If you find too much resistance to your effort to remove the rod, leave it in place and clean the tank as best as you can. If it has broken off and you don't have time to disassemble the entire tank and rotary valve to get out the stub, lay a brick on top of the holding wheel when you put the unit back together to keep it and the grids in place. Then come back and fix it when you have time.

5. **Manifold Removal** Reach in the tank and remove the manifold. It rests just inside the rim of the rotary valve; it is not threaded in place.

6. **Cleaning** Hose out the inside of the tank, the manifold, and the holding wheel. Hose off the grids (Fig. 5-9B). You might need to scrub them lightly with a soft bristle brush to loosen the grime. If the grids are still dirty, soak them in a garbage can of water, trisodium phosphate (1 cup per 5 gallons of water or 250 milliliters per 19 liters), and muriatic acid (1 cup per 5 gallons) (acid alone will not clean grids because it does not affect the oil). After 30 minutes, try scrubbing them clean again. Don't use soap—you won't get it all out, no matter how well you rinse the grids, and when you start up the circulation again you'll have soap suds in the pool.

7. **Reassemble Rod and Manifold** Inspect the manifold for chips or cracks. DE and dirt will go through such openings and back into your pool. Cracks can be glued. If chunks of plastic are missing, buy a new manifold—they're only about $30. Particularly inspect the joint between the top and bottom halves of the manifold. Where these two parts are glued together, they often start to separate. Replace the manifold as you took it out. Reinstall the center rod.

8. **Reassemble Grids and Retainer** Carefully inspect the grids before putting them back inside. Look for worn or torn fabric, cracked necks on the nipples, or grids where the plastic frame has collapsed inside the fabric. Replace any severely damaged grids. When you reinstall the grids, notice that inside each hole in the manifold is a small nipple and on the outside of each grid nipple is a small notch. By lining up the nipple and notch as you reinsert each grid, the grids will go back as intended.

Now lay the retainer over the tops of the grids and spin it around until it finds its place holding down and separating the grids. Screw on the wing nut and washer that holds down the retainer holding wheel.

9. **O-Ring Reassembly** Getting the lid back on can be as tough as getting it off. Make sure the O-ring on the tank is free of gouges and has not stretched. If it is loose, soak it for 15 minutes in ice water and it might shrink back to a good fit. If not, replace it. Apply tile soap as a lubricant to make it slide on easier (or silicone lube if you can afford it) to the inside of the lid around the edge that will meet the O-ring. Don't use Vaseline or petroleum-based lubricants because these will corrode the O-ring material. Don't use that green slime called Aqua Lube—it sticks to everything and comes off of nothing.

10. **Lid Reassembly** Now close the tank drain, turn the backwash valve to normal filtration, and turn on the pump. Let the tank fill with water. Turn off the pump and turn the valve to backwash. The water will drain out, sucking the lid down. Don't be afraid to help it along by getting on top of the lid. Your weight will finish the job. Be careful not to hit the pressure gauge assembly—they snap off very easily.

11. **Clamp Reassembly** Replace the clamping ring, return the valve to normal filtration, and start the pump/motor. Open the air relief valve and purge the air until water spurts out the valve.

12. **Add DE** Never run a DE filter without DE, even for a short time. Dirt will clog the bare grids. Remember, it's not the grids, but the DE that does the actual filtering. The label on the filter will tell you how much DE to add, or refer to the table on the bag of DE. It tells you how many pounds of DE to add per square foot of filter area. As a convenient scoop, use a 1-pound (½-kilogram) coffee can; but remember that because DE is so light and powdery, a 1-pound coffee can holds only ½ pound (226 grams) of DE. One pound of DE covers 10 square feet (1 square meter) of grids.

DE is added to the system through the skimmer. Do not dump it in all at once. It will form in clumps at the first restricted area, like a plumbing elbow or the inlet on the filter tank. Sprinkle in

TRICKS OF THE TRADE: DE

- A good way to add DE to prevent clumping is in a slurry. On large commercial installations that require large amounts of DE, there is actually a slurry pit, but you can use a bucket. The concept is that you thoroughly mix the DE in water (achieving that suspension I spoke of) before pouring the solution into the skimmer.

- Most pools have skimmers where you can add DE. But what if there isn't one? Again, make a slurry in a bucket. After cleaning the filter, take the lid off the pump strainer pot and turn on the pump. Add the slurry to the strainer pot, followed by clear water (have a hose handy) to make sure all the DE gets to the filter and completely and evenly coats the grids. Turn off the pump, fill the pot with water, replace the lid, and reprime the system. Again, remember to let the air out of the tank.

one can of DE at a time, mixing it in the skimmer water with your hand. It should appear to be dissolving in the water. In fact, it is not dissolving, just freely suspending itself in the water, but this will keep it from clumping.

Add one can and wait about a minute. If you have any gaps in the manifold or holes in the grids or you didn't assemble the unit correctly, DE will get through these areas and flow back into the pool. If that happens, it is better to have one can of DE flowing back into the pool rather than all 10 or 15. On most start-ups you will see a little milky residual entering the pool from any DE or dirt that settled in the pipes during cleaning. This is normal, but if you see great clouds of DE returning to the pool, shut the system down and take the filter apart. You missed something.

I must emphasize this problem of adding DE too fast (or using too much—follow the DE package directions). Comedian Rich Little's system was sluggish and the filter pressure unusually low. I cleaned the pump strainer, skimmers, blew water through the suction lines with a drain-flush bag, and cleaned the filter twice . . . and no luck. I took apart the pump to make sure the impeller was clean and operating properly. I checked the power supply, thinking a bad breaker was perhaps delivering low voltage and making the motor work too slowly. No luck.

Finally, after hours of hunting, I emptied the filter tank and fed hose water directly into the pump and watched it trickle into the tank. I knew the obstruction was in the plumbing between the pump and the rotary valve. I cut open the plumbing and sure enough, at one of the 90-degree elbows I found clumps of DE. The pressure had crystallized it and made it rock hard. I had to replace that section of plumbing and later in the shop I took a hammer and chisel to the DE to see just how hard it had become.

SAND FILTERS

RATING: EASY

Sand filters are designed to use #20 silica sand, a specific size and quality of sand. Larger sand will not filter fine particles from the water and finer sand is pushed through the slits in the laterals, clogging them.

> **TRICKS OF THE TRADE: ALUM**
>
> If channeling is a problem in your sand filter because of hard water or pool chemistry (which speeds up calcification of the sand), introduce aluminum sulfate (alum) through the skimmer just like you would add DE to help prevent this problem. Use the amounts recommended on the bag, but usually about ½ cup per 3 square feet (125 milliliters per 0.3 square meter) of filter size.

Sand filters do need regular backwashing and, unlike DE backwashing, it is effective. Although wet sand is heavy and lies on the bottom, when the tank is full of circulating water the sand is suspended in the tank. In fact, you can reach into a tank of sand and water and get your hand all the way to the bottom of the tank. Try that without water in the tank. Your hand will push about 6 inches (15 centimeters) into the thick sand.

Backwashing: Most rotary valves have the steps printed right on them, and they are very simple.

1. **Prepare** Turn off the pump. Rotate the valve to Backwash. Roll out your backwash hose or make sure the waste drain is open.

2. **Flush** Turn on the pump and watch the outgoing water through the sight glass. It will appear clean, then dirty, then very dirty, then it will slowly clear. When it is reasonably clear, turn off the pump and rotate the valve to Rinse.

3. **Rinse** Turn the pump back on and run the rinse cycle for about 30 seconds to clear any dirt from the plumbing. Turn off the pump, rotate the valve back to Filter, and restart the pump for normal filtration.

Backwash as often as necessary. When the filter gauge reads 10 psi (700 millibars) more than when the filter is clean, it is usually time to backwash. A better clue is when dirt is returning to the pool or when vacuuming suction is poor.

When backwashing, be sure there is enough water in the pool to supply the volume that will end up down the drain. It is usually a good idea to add water to the pool each time you backwash.

Teardown: Twice per year, I recommend opening the filter. Sand under pressure and with the constant use of pool chemicals or dissolving pool plaster will calcify, clump, and become rock-like over time. Passages are created through or around these clumps, but less and less water is actually filtering through the sand and more is passing around it. This is called *channeling*. To correct or avoid this problem, regular teardown is the answer.

1. **Shutdown** Turn off the pump. Disconnect the multiport valve plumbing by backing off the threaded union collars. Some valves are threaded into the body of the tank, others are bolted on. Remove the valve.

2. **Flush** Some sand filters have a large basket just inside the tank. Remove this and clean it out. The sand is now exposed. Push a garden hose into the tank and flush the sand. As noted previously, it will float and suspend in the water. Use a broom handle to bust up clumps. As the water fills the tank, it will overflow, flushing out dirt and debris. Be careful not to hit the laterals on the bottom of the tank because they are fragile and break easily.

3. **Reassemble** When the sand is completely free and suspended in the water, not clumped, turn off the water and replace the basket (if any), multiport valve, and plumbing. Backwash briefly to remove any dirt that was dislodged by this process but not yet flushed out.

This teardown process also allows you to check to see if the regular backwashing has flushed out too much sand. You might need to add some fresh sand. Most sand filters need to be filled about two-thirds with sand and have one-third freeboard. Backwash after adding any new sand to remove dust and impurities from the new sand.

Replacing Sand: Every few years you need to replace the sand completely because erosion from years of water passing over each grain makes them round instead of faceted and rough. Smooth sand does not

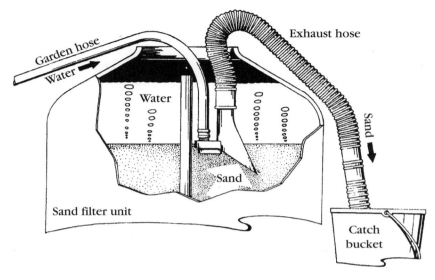

FIGURE 5-10 **Sand filter vacuum.** *Lass Enterprises, Inc.*

catch and trap dirt as efficiently and it slowly erodes to a smaller size than the original #20 silica, allowing it to clog laterals and pass into the pool. Some manufacturers suggest adding a few inches of gravel over the laterals first. This keeps the sand separated from the laterals so the sand cannot clog them. To replace sand, or add sand to a new installation:

1. **Remove**   Open the filter as described previously. Remove the old sand by scooping it out with your hands or with a sand-vac (Fig. 5-10).

2. **Add Water**   Fill the bottom third of the tank with water to cushion the impact of the sand on the laterals.

3. **Add Sand**   Slowly pour the sand into the filter, being careful of the laterals. Fill sand to about two-thirds of the tank. Reassemble the filter parts and backwash to remove dust and impurities from the new sand, then filter as normal.

CARTRIDGE FILTERS

RATING: EASY

Cleaning a cartridge filter is perhaps easiest of all.

1. **Shutdown**   Turn off the pump. Remove the retaining band and lift the filter tank or lid from the base. Remove the cartridge.

FAQS: FILTERS

Which is better: sand, cartridge, or DE filters?

- Each filter type will keep a pool clean. The key to success is proper sizing and regular cleaning. The best type is the one that you are most likely to keep clean, so a cartridge filter may be the best choice because of the ease of maintenance. That said, if you have a large pool, a cartridge filter may not be practical, so choose the filter that fits your pool—and clean it often!

If I backwash, do I also need to tear down and clean my DE or sand filter?

- Backwashing of DE filters should be used as a temporary cleaning measure when complete teardown and cleaning is impractical. Sand filters respond well to backwashing, and since there is no DE to add back, there is no potential for error that may lead to a dirty pool.

How often should I backwash?

- The schedule for backwashing and teardown of a filter is a factor of how much the pool is used and how dirty it gets in normal service. Typically, DE filters should be torn down and cleaned fully at least six times per year. Unless your pool gets very dirty, you won't need to backwash it. Sand filters can be backwashed once a month and torn down twice a year.

Why is there DE or dirt passing back into the pool?

- Dirt passes back into the pool because of torn grids or improper reassembly after cleaning. Take the filter apart and check all components; then reassemble it carefully and try again.

When replacing a filter, should I buy a bigger one?

- Bigger is not always better. If you find your filter needs cleaning more than once a month, it may be undersized. Consult a pool professional to get a new filter of the proper size—a filter that's too large for your pump will not fill with water and therefore will not give you the improved results you expected.

2. **Clean Cartridge** Light debris can simply be hosed off, but examine inside the pleats of the cartridge. Dirt and oil have a way of accumulating between these pleats. Never acid wash a cartridge. Acid alone can cause organic material to harden in the web of the fabric, effectively making it impervious to water. Soak the cartridge in a garbage can of water with trisodium phosphate (1 cup per 5 gallons or 250 milliliters per 19 liters) and muriatic acid (1 cup per

5 gallons). About an hour should do it. Remove the cartridge and scrub it clean in fresh water. Don't use soap. No matter how well you rinse, some residue will remain and you will end up with suds in your water.

3. **Reassemble** Reassemble the filter and resume normal circulation.

Buy a spare cartridge so you can install it while soaking the dirty one. Replace cartridges when they won't come clean, when the webbing of the fabric appears shiny and closed, or when the fabric has begun to deteriorate or tear.

Skidpacks

Now that you understand pumps, motors, filters and their related plumbing, you understand all aspects of the above-ground pool equipment package, or *skidpack*. The only additional element is a stand, which mounts these components together, called a *skid*.

Figure 5-11 shows a typical skidpack, which is plumbed to the pool with rigid PVC plumbing and powered by household current from the skidpack's standard cord. Note how the pump and filter are connected with a threaded union, making disassembly simple. The only drawback of the skidpack is its proximity of the filter to the motor. Routine cleaning or leaks can result in water flooding the motor, with obvious consequences. When working with the filter of a skidpack, always unplug the unit and be especially cautious of water flowing onto the motor. If it does, allow the motor to dry thoroughly before plugging it back in.

Figure 5-12 shows the parts breakdown of a typical skidpack, this one featuring a

FIGURE 5-11 Typical skidpack.

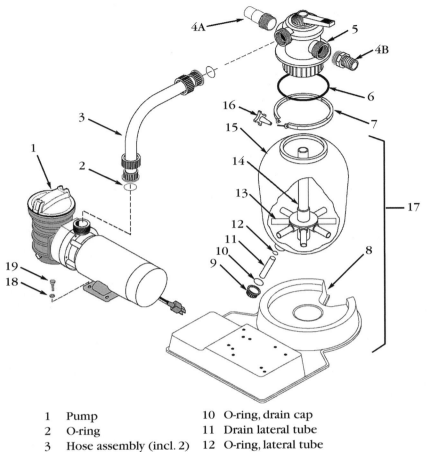

1	Pump
2	O-ring
3	Hose assembly (incl. 2)
4A	Hose adapter
4B	Sight glass
5	Multiport valve
6	O-ring, tank flange
7	V-clamp with knob
8	Pedestal platform
9	Drain cap
10	O-ring, drain cap
11	Drain lateral tube
12	O-ring, lateral tube
13	Lateral tube
14	Collector hub assembly
15	Tank assembly
16	Clamp knob
17	Filter tank assembly (incl. 8—15)
18	Washer $1/4$"
19	Screw $1/4$"-20 $\times$ $3/4$"

FIGURE 5-12 **Typical skidpack components.** *Sta-Rite Industries.*

sand filter. The skid is molded plastic, designed to accommodate the shapes of the equipment bases. The motor and filter simply bolt onto the skid. Although most skidpacks consist of just a pump, motor, and filter, many are designed for above-ground pools with optional equipment such as a chlorinator, heater, or timeclock.

Remember that a skidpack is powered by electricity in close proximity to water, so only plug it into a waterproof household electrical outlet that is fitted with a ground fault circuit interrupter (Fig. 5-13). Also be careful with cords laid across any part of the yard where people are likely to trip over them, lawnmowers may run over them, or sprinklers may soak them. If you need a longer cord, buy one and install it rather than relying on an extension cord.

FIGURE 5-13 Electrical outlet with ground fault circuit interrupter.

Heaters

The basic principle of the pool heater (Fig. 6-1) is simple. A gas burner tray creates heat. Heat rises through the cabinet of the heater, raising the temperature of the water that is passing through the serpentine coils above.

Gas-Fueled Heaters

Figure 6-2 shows a typical gas-fueled heater (those that use natural or propane gas as the heating fuel). The water passes in one port of the front water header (item 23), then through the nine heat exchanger tubes (item 45). The water reaches the rear header (item 24) and is returned through other exchanger tubes to the front header and out the other port.

Most modern exchangers are four-pass units, meaning the water goes through at least four of the tubes, picking up 6° to 9°F (3° to 5°C) on each pass, before exiting the heater. Generally, these are self-cleaning unless extreme calcium (scale) is present in the water.

The heat exchanger tubes are made of copper, which conducts heat very efficiently. The tubes have fins (about eight per inch or 2.5 centimeters) to absorb heat even more efficiently and are topped with sheet metal baffles (item 53) to retain the heat. The heat rising from the burner tray (item 55) is effectively transferred to the water in the exchanger because of the excellent conductivity of copper; however, improper

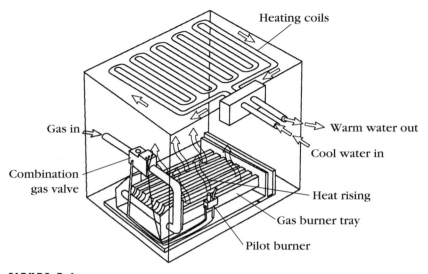

Heating coils

Gas in

Combination
gas valve

Warm water out

Cool water in

Heat rising

Gas burner tray

Pilot burner

FIGURE 6-1 The concept of the pool heater.

water chemistry can easily attack this soft metal and dissolve it into the water. More on that later.

Notice that there is a flow control assembly on the front header (items 38–42). This spring-loaded valve is pressure sensitive, designed to mix cool incoming water with hot outgoing water to keep the temperature in the exchanger from becoming excessive. This design keeps the outgoing water no more than 10° to 25°F (5° to 10°C) (depending on manufacturer) above the temperature of the incoming water to prevent condensation and other problems that greater differentials would create. Temperature control is achieved by flow regulation rather than direct temperature regulation. Maintaining a constant flow through the heat exchanger results in a constant water temperature.

When reaching temperatures over 115°F (46°C), water breaks down, allowing minerals suspended in it to deposit in the heat exchanger. Also, water is designed to flow through the unit at no more than 100 gpm (378 liters per minute) with 1½-inch (40-millimeter) plumbing or 125 gpm (473 liters per minute) with 2-inch (50-millimeter) plumbing. Above that, a manual bypass valve is installed.

The other major component of the gas-fueled heater is the burner tray. This entire assembly can be disconnected from the cabinet and pulled out for maintenance or inspection. Depending on the size of the heater, there will be 6 to 16 burners (item 6), the last one on the right

TRICKS OF THE TRADE: HEATER SAFETY

Heaters are unquestionably the most potentially dangerous component of the pool equipment group. They combine water under pressure and heat, gas or other combustible fuel, and electricity. The point is simply that whatever care you exercise normally must be doubled when working with heaters. Therefore, I have a simple safety checklist for working around heaters.

- Never bypass a safety control and walk away. Jumping controls (discussed below) is a good way to troubleshoot, but do not operate the unit this way. Always remove your jumpers after troubleshooting.

- Never repair a safety control or combination gas valve. Replace it. You will notice that your supply house doesn't even sell parts for gas valves. They should never be repaired, because future failure could be catastrophic.

- Never hit a gas valve—it might come on, but it might stay on.

- Keep wiring away from hot areas and the sharp metal edges of the heater.

- Disable the heater and tape a shutdown notice on the unit until repairs are made. *You can be held liable* if you are the last person to work on a heater and it causes damage or injury by firing incorrectly or before repairs have been made.

- When jumping a safety control or otherwise trying to fire a heater that will not come on, keep your face and body away from the burner tray, where flashback might occur. It might be awkward to squat alongside a heater and jump a control through the opening in the front, but awkward is better than burned. Double that warning with LP-fueled units.

having a pilot (item 5) mounted on it. Individual burners can be removed for replacement. The combination gas valve (item 3) regulates the flow of gas to the burner tray and pilot and is itself regulated by the control circuit.

Gas-fueled heaters are divided into two categories based on the method of ignition.

The Millivolt or Standing Pilot Heater

As the name implies, the standing pilot system of ignition uses a pilot light (burner) that is always burning or is ignited by a spark from a piezo ignition unit like many gas barbecues. The heat of the pilot is converted into a small amount of electricity (0.75 volt or 750 millivolts)

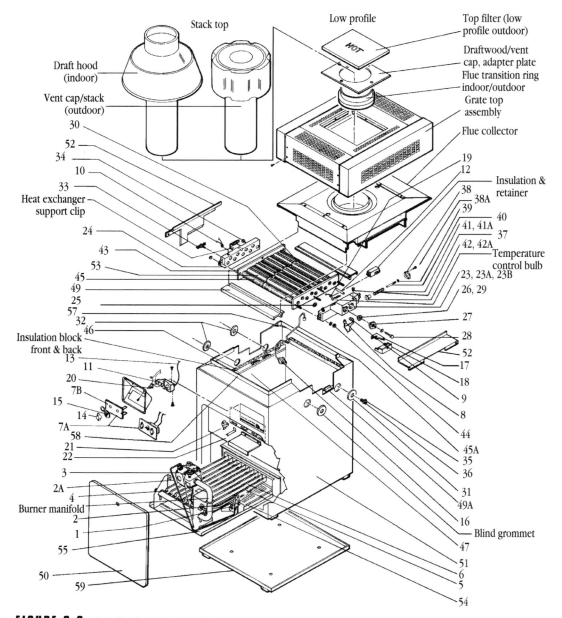

FIGURE 6-2 Detail of a typical millivolt, standing pilot heater. *Image courtesy of Jandy.*

Key No.

1 Pilot generator assembly
2 Visoflame lighter tube
2A Pilot tube
3 Automatic gas valve
4 Burner orifice
5 Burner w/pilot bracket
6 Burners
7A Plate assembly
7B Thermostat dial
8 High-limit switch (135°F.)
9 High-limit switch (150°F.)
10 Redundant limit
11 Temperature control
12 Protective sleeve, bulb
13 Wire harness
14 Thermostat knob
15 Temp-lok
16 Pressure switch
17 High-limit switch retainer clip
18 High-limit switch cover
19 O-ring
20 On-off switch
21 Fusible link
22 Fusible link bracket
23, 23A, 23B Front water header
24 Rear water header
25 Header gasket
26, 29 Flange packing collar
w/copper sleeve
27 Water header flange
28 Water header flange bolt

30 Heat exchanger baffle retainer
31 Drain grommet
32 Drain grommets
33 Drain valve
34 Drain plug
35 Drain valve
36 Bushing, drain valve
37 Brass plug
38 Flow control cap
38A, 43 Bolt & washer
39 Flow control gasket
40 Flow control shaft
41, 41A Front control spring
42, 42A Flow control disc
43 Bolt, front & rear header
44 Header nut, 3/8" hex
45 Heat exchanger
45A Syphon loop
46 Fiberglass blanket
47 Insulation block, side
49 Insulation block cover, front & back
49A Insulation block cover, end
50 Door
51 Jacket assembly
52 Gap closure
53 Heat exchanger baffle
54 Burner tray shelf
55 Burner tray assembly
57 Rear deflector
58 Lower deflector
59 Noncombustible floor base (optional)

FIGURE 6-2 (*Continued*)

by a thermocouple which in turn powers the control circuit. The positive and negative wires of the thermocouple (also called the pilot generator) are connected to a circuit board on the main gas valve.

If your heater does not use a piezo ignition, when lighting the pilot, it is necessary to hold down the gas control knob to maintain a flow of gas to the pilot. When the heat has generated enough electricity (usually a minimum of 200 millivolts) the pilot will remain lit without holding the gas control knob down. The positive side of the thermocouple also begins the electrical flow for the control circuit. When electricity has passed through the entire control circuit, the main gas valve opens and floods the burner tray with gas which is ignited by the pilot.

The Control Circuit

The control circuit is a series of safety switches—devices that test for various conditions in the heater to be correct before allowing the electrical current to pass on to the main gas valve and fire up the unit. Figure 6-3 shows a millivolt control circuit; Fig. 6-4, the parts of an electronic control circuit. Following the flow of electricity (not all manufacturers follow the same routing of their control devices, but they all include the same devices), a control circuit includes the following items.

FUSIBLE LINK

The fusible, or fuse, link (Fig. 6-2, item 21) is a simple heat-sensitive device located on a ceramic holder near the front of the gas burner tray.

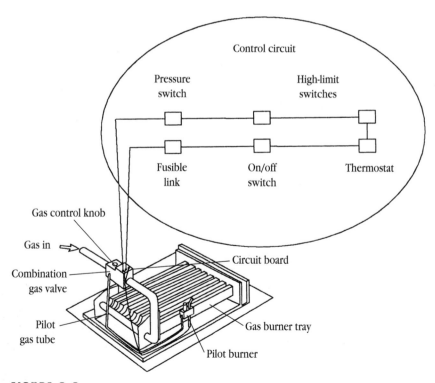

FIGURE 6-3 The control circuit.

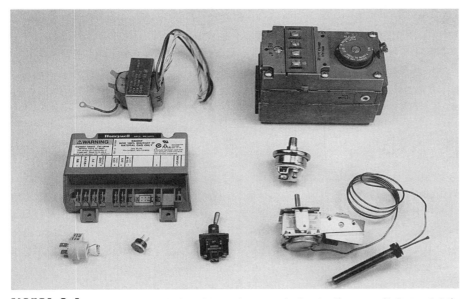

FIGURE 6-4 Components of the electronic control circuit. Top row (left to right): transformer, automatic combination gas valve. Middle row: intermittent ignition device (IID), pressure switch. Bottom row: fusible link, high-limit switch, on/off switch, mechanical thermostat.

If the heat becomes too intense, the link melts and the circuit is broken. This would most commonly occur when debris (such as a rodent's nest or leaves) is burning on the tray or if part of a burner has rusted out causing high flames. Other causes are improper venting (allows excessive heat build-up in the tray area), extremely windy conditions, or low gas pressure causing the burner tray flame to roll out toward the link.

Figure 6-5 diagrams a fusible link. A wax pellet, designed to melt at certain high temperatures, melts and allows the spring to break its normal contact, thus cutting power in the circuit. My early experience was that these links often needed to be replaced even when the heater was new or operating normally; however, current engineering has overcome the oversensitivity of these devices. When the fusible link burns out, it pays to examine the other components of the heater or the installation to find the cause.

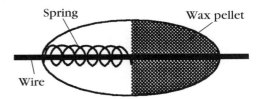

FIGURE 6-5 Detail of the fusible link.

Improper venting is not the only cause of overheating. Rusted components, improper installation, low gas pressure, and insect or rodent nests more frequently cause the problem, so the fusible link is one more safety device in a component of your pool equipment that needs a lot of safety.

Not all manufacturers included a fusible link in their control circuit when they were first introduced by Teledyne Laars. Today, virtually all heaters use one.

ON/OFF SWITCH

As the name implies, the on/off switch (Fig. 6-2, item 20) is usually a simple, small toggle-type switch on the face of the heater next to the thermostat control. Sometimes, particularly on older Teledyne Laars models, the switch is located on the side of the heater in a separate metal box that also contains the thermostat. In older Raypak models, the switch and thermostat might be located on the side, mounted directly through the sheet metal side of the heater cabinet.

Often the switch is remotely located so the user can switch the unit on and off from a more convenient location than where the equipment itself is located. Manufacturers recommend that a remote on/off switch for a millivolt heater be located no more than 20 to 25 feet (6 to 7 meters) from the heater. This is because with less than 0.75 volt passing through the control circuit, any loss of voltage from running along extended wiring means that there might not be enough electricity left to power the gas valve when the circuit is completed.

Also, as the thermocouple wears out and the initial electricity generated decreases, the chance that there won't be enough power becomes very real. Therefore, I suggest from experience that remote switches be located no more than 10 feet (3 meters) from the heater and that they be run through heavily insulated wiring to avoid heat loss. Better yet, run a remote switch off a relay so the control circuit wiring does not have to be extended at all.

If the heater has two thermostats, the switch has three positions: high, off, and low. Why would you have more than one thermostat? Let's say you have a pool and spa, both operated by the same equipment. You can set one thermostat for your desired pool temperature and the other for your desired spa temperature. Then, instead of having to reset the thermostat when you run the spa, you simply flick the switch to read the second thermostat. Many manufacturers now sell only dual thermostat heaters because it costs supply houses too much to stock heaters of both kinds when the difference is only a few bucks for one additional thermostat. You won't need the extra switch for your above-ground pool, but that's why it's there.

It is also worth noting that if the switch has been remotely located or duplicated in a remote on/off system, the factory-installed unit might be left in place but not be operative. Factory technicians will place a sticker above such a nonfunctioning switch to alert future users or technicians. However, they are often lost or not used by pool builders or other technicians and as a result, you might be confused when troubleshooting the heater.

THERMOSTATS

Thermostats (Fig. 6-2, item 11), also called temperature controls, fall into two categories: mechanical and electronic. The mechanical thermostat is a rheostat dial connected to a metal tube that ends in a slender metal bulb. The tube is filled with oil and the bulb is inserted in either a wet or dry location where it can sense the temperature of the water entering the heater. These thermostats are precisely calibrated, but many installation factors affect the temperature results. In other words, setting the dial at a certain point might result in 80°F (26°C) water in one pool while the exact same setting might result in 85°F (29°C) water in another pool. Therefore, pool heater thermostats generally are color-coded around the face of the dial, showing blue at one end for cool and red at the other end for hot. Settings in between are by trial and error to achieve desired results.

Usually, as shipped from the factory, thermostats will not allow water in the pool to exceed 103° to 105°F (39° to 40°C), although they can be set higher. Also, they do not generally register water cooler than 60°F (15°C), so if the water is cooler than that you might turn the thermostat all the way down and the heater will continue to burn. Therefore, the only way to be sure a heater is off is to use the on/off switch (or turn off the gas).

The electronic thermostat uses an electronic temperature sensor that feeds information to a solid-state control board. These are more precise than mechanical types; however, because of the same factors noted previously they are also not given specific temperatures, but rather the cool to hot, blue to red graduated dials for settings. Some manufacturers of spa controls make specifically calibrated digital thermostats, but my experience is that no matter what the readout says, the actual temperature will vary greatly.

HIGH-LIMIT SWITCHES

High-limit switches (Fig. 6-2, items 8, 9, and 10) are small, bimetal switches designed to maintain a connection in the circuit as long as their temperature does not exceed a predesigned limit, usually 120° to 150°F (49° to 65°C). The protection value is similar to the fusible link and often two are installed in the circuit, one after the other, for safety and to keep the heater performing as designed.

The first high-limit switch is usually a 135°F (57°C) switch, and the other is a 150°F (65°C) switch. Where the fusible link detects excessive air temperatures, the high-limit switch detects excessive water temperatures. They are mounted in dry wells in the heat exchanger header. Sometimes a third switch, called the *redundant high-limit*, is mounted on the opposite side of the heat exchanger for added safety.

PRESSURE SWITCH

The pressure switch (Fig. 6-2, items 16 and 45A) is a simple switching device at the end of a hollow metal tube (siphon loop). The tube is connected to the header so that water flows to the switch. If there is inadequate water flow in the header there will not be enough resulting pressure to close the switch. Thus, the circuit will be broken and the heater will shut down.

Although preset by the factory (usually for 2 psi or 138 millibars), most pressure switches can be adjusted to compensate for abnormal pressures caused by the heater being located unusually high above or below the water level of the pool or spa.

AUTOMATIC GAS VALVE

The automatic gas valve (Fig. 6-2, item 3) is often called the *combination gas valve* because it combines a separately activated pilot gas valve with a main burner tray gas valve (and sometimes a separate pilot-lighting gas line combined with the pilot gas valve). After the circuit is complete, the electricity activates the main gas valve which opens, flooding the burner tray. The gas is ignited by the pilot and the heater burns until the control circuit is broken at any point, such as when the desired temperature is reached and the thermostat switch opens, if the on/off switch is turned to off, if the pressure drops (such as when the time clock turns off the pump/motor) and the pressure switch opens and breaks the circuit.

There are several different designs of automatic gas valve, but all have aspects in common. Figure 6-6 shows a typical valve design. For millivolt-powered units, the terminal board will have three terminals (or four, where terminals 2 and 3 are connected by a common connection or fusible link). Terminals 1 and 2 are the neutral and electric hot (positive) lines from the pilot generator to start the control circuit. When the circuit is complete, power arrives at terminal 3 (or 4) which opens the main gas valve.

On 25-volt units, there is a pair of terminals to power open the pilot valve and to return the current to the common (or neutral) line of the

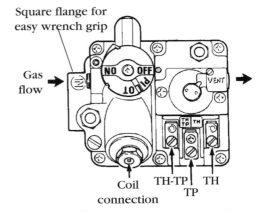

FIGURE 6-6 **The automatic (combination) gas valve.** *Image courtesy of Jandy.*

intermittent ignition device (IID) (detailed later). Another pair of terminals power open the main gas valve and return the current to the common (or neutral) line of the IID.

If you are unsure how a valve should be wired, look for markings near each terminal. They are often marked *PP* meaning powerpile (the pilot generator) or *TP* meaning thermopile (which are the same thing and, just to confuse you, they're also called thermocouple, but I have never seen *TC* on a gas valve terminal); *TH* meaning thermostat or other connection to the control circuit; and *TR* meaning transformer.

The gas plumbing of the automatic gas valve is self-explanatory. The large opening (½ or ¾ inch or 13 or 19 millimeters) on one end, with an arrow pointing inward, is the gas supply from the meter. Note that it has a small screen to filter out impurities in the gas, like rust flakes from the pipe. The hole on the opposite end feeds gas to the main burner. The small threaded opening is for the pilot tube and a similar hole is for testing gas pressure. These are clearly marked. Older Teledyne Laars heaters employ an additional small tube to assist in lighting the pilot (see visoflame tube).

Automatic gas valves are clearly marked with their electrical specifications, model numbers, and most important, Natural Gas or Propane. Black components or markings usually indicate propane.

All combination gas valves have on/off knobs. On 25-volt units, the knob is only on or off. With standing pilot units, there is an added position for *pilot* when lighting the pilot. As a positive safety measure in most, you are required to push the knob down while turning.

Electronic Ignition Heaters

When a heater (Fig. 6-7) with electronic ignition is turned on, an electronic spark ignites the pilot which in turn ignites the gas burner tray in the same manner as described previously. In all other respects, these heaters operate the same way as those already discussed. Where the control circuit on the

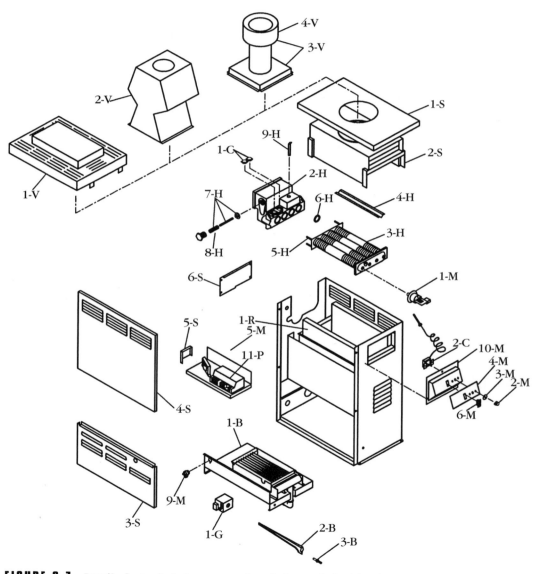

FIGURE 6-7 Detail of a typical above-ground pool electronic ignition heater. *Raypak, Inc.*

standing pilot heater is powered by millivolts, the electronic ignition heater is controlled by the same kind of circuit but is powered by 25 volts ac (Fig. 6-8). Regular line current (at 120 or 240 volts) is brought into the heater and connected to a transformer that reduces the current to 25 volts.

This voltage is first routed into an electronic switching device called the IID (intermittent ignition device), which acts as a pathway to and from the control circuit. From here the current follows the same path through the same control circuit switches as described previously.

When the circuit is completed the current returns to the IID, which sends a charge along a special wire to the pilot ignition electrode creating a spark that ignites the pilot flame. The IID simultaneously sends current to the gas valve to open the pilot gas line.

When the pilot is lit, it is sensed by the IID through the pilot ignition wire. This information allows the IID to open the gas line to the burner tray, which is flooded with gas ignited by the pilot.

Natural versus Propane Gas

The differences between heaters using natural gas and those using propane gas are nominal. Most manufacturers make propane heaters

Key No.

1-B	Burner tray with burners (sea level)	4-M	Dial plate
2-B	Burner	5-M	Transformer (IID unit)
3-B	Burner orifice (sea level)	6-M	Rocker switch (SPST)
1-C	High limit 135°F	9-M	Thermal fuse
2-C	Thermostat control	10-M	Bezel (less label)
1-G	Combination valve	11-P	Ignition control IID
2-H	Inlet/outlet header	1-R	Refractory block kit
3-H	Tube bundle	1-S	Jacket top
4-H	Baffle kit	2-S	Flue collector
5-H	Carriage bolt kit	3-S	Door assembly
6-H	Header gasket (4)	4-S	Upper jacket control panel
7-H	Bypass valve	5-S	Wiring box
8-H	Bypass spring	6-S	Access panel (inlet/outlet end)
9-H	Well retaining clip	1-V	Stackless top (outdoor)
1-M	Pressure switch	2-V	Drafthood (indoor)
2-M	Thermostat knob	3-V	Outdoor stack with adapter (outdoor)
3-M	Knob stop	4-V	Outdoor stack

FIGURE 6-7 (*Continued*)

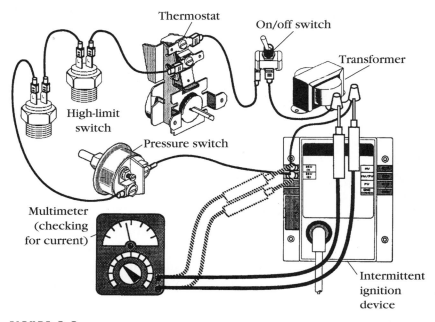

FIGURE 6-8 Electronic ignition heater control circuit (with manual thermostat). *Raypak, Inc.*

in standing pilot/millivolt models only. Because of different operating pressures, the gas valve is slightly different (although it looks the same as a natural gas model), as are the pilot light and the burner tray orifices. The gas valve is clearly labeled *Propane*. The heater case, control circuit, and heat exchanger are all the same as for a natural gas model.

Natural gas is lighter than air and will dissipate somewhat if the burner tray is flooded with gas but not ignited for some reason. Similarly, the odor added to natural gas will be detected if you are working nearby as the gas floats out and upward. Make no mistake, this is still a serious situation and explosions can occur.

Solar-Fueled Heaters

Solar heating systems are becoming more user friendly every year, especially for above-ground pools. Panels are made of lighter materials than just a few years ago, including some made from doormat-like rubber that simply tacks onto a roof or hillside in large, flexible sheets. Most manufacturers of this technology or traditional panels sell complete kits—systems composed of the necessary panels, controls, and installation instructions. The most common solar heating systems for above-ground pools are black plastic panels that lie on the ground or lean against the pool wall. Water circulates through them, picking up heat and transferring it to the pool. The only regulation is to run the circulation system more or fewer hours (for more or less heat) or to use bypass valves that allow you to cut off the solar panels from the pool plumbing. More complex systems are also used for larger above-ground pools, including those pools that already have a gas-fueled heater.

Refer back to Fig. 3-6, which shows a typical solar installation. The concept to understand here is that the water should go

> ## TRICKS OF THE TRADE: PROPANE SAFETY
>
> **Propane gas is heavier than air and if it floods the burner tray without being ignited it tends to sit on the bottom of the heater. Because it remains undissipated and because you are less likely to smell it because it is not floating out and upward, if it does suddenly ignite it will do so with violent, explosive force. Rarely is the heater itself damaged—the explosion takes the line of least resistance, which is out through the open front panel. Never position your face in front of the opening to try to learn why the heater hasn't fired. Remember to follow your safety checklist, and treat propane with great respect.**

through the solar panels before it passes through the heater. In this way, whatever heat can be gained from the sun is obtained first, then the gas heater adds additional heat if desired. This type of solar heating system can be controlled by time clocks and/or thermostats because, in summer, the panels might add too much warmth to the water and some means of regulation is needed. Also, they typically have simple on/off toggle switches to completely disable the system.

Since solar heating is essentially a plunbing job, it is described in more detail in Chap. 3.

Makes and Models

You'll learn more about the sizes of heaters and their functions in the next section. For now you need to know that heater models are based on their size as expressed in output of heat (measured in Btus). Each manufacturer produces models of similar size; for example, 50,000, 125,000, 175,000, 250,000, 325,000, and 400,000 Btu. After this range, you enter the realm of commercial heaters.

Because of frequent changes in engineering and design, what I say in the following section about a particular manufacturer might be true about some models and not others. I have tried to make comments based on the heaters I find most commonly in use today.

Heaters designed for above-ground pools are identical to those made for inground pools, although electronic ignition heaters are sometimes designed to be plugged directly into household electrical outlets, rather than being hard-wired into breaker panels or time clocks, as you might for inground pool equipment. Some models are also equipped with a blower to force more hot air over the heat exchanger, resulting in a more rapid rise in temperature from a lower-rated (Btus) heater.

Selection

Beyond manufacturer preferences or price, there are two basic parameters to consider in selecting a heater: sizing and cost of operation.

Cost of Operation

The cost of operating a heater is simple to figure out if you know what your customer pays for a therm of gas. A *therm*, the unit of measurement you read on the gas bill, is 100,000 Btu/hour of heat. My last gas bill showed I pay about 50 cents per therm. The heater model tells you how many Btus per hour your heater uses. Divide that by 100,000 to tell you how many therms per hour it is. Next, determine how many hours of operation are needed to bring the temperature up to the desired level.

Let's look at a simple example. You need at least a 135,000 output Btu heater, so you buy a 250,000 input Btu model. You decide it will run 8 hours a day. Using the facts you have, here is your operating cost:

250,000 Btu ÷ 100,000 Btu/h = 2.5 therms/h

2.5 therms/h × 8 h/day = 20 therms/day used to run the heater

20 therms/day × 50¢/therm = $10/day

So how much will it cost to keep a standing pilot burning. Well, it uses between 1200 and 1800 Btu per hour. Figure it out from there. By the way, the temperature of that little flame is over 1100°F (593°C), so when removing a pilot assembly for repair, don't grab one that has recently been lit.

Pools

	10°F/5°C	15°F/7°C	20°F/10°C	25°F/13°C	30°F/16°C
200 sf/18 sm	21,000 Btu	32,000 Btu	42,000 Btu	53,000 Btu	63,000 Btu
400 sf/36 sm	42,000 Btu	63,000 Btu	84,000 Btu	105,000 Btu	126,000 Btu
600 sf/54 sm	63,000 Btu	95,000 Btu	126,000 Btu	157,000 Btu	189,000 Btu
800 sf/72 sm	84,000 Btu	126,000 Btu	168,000 Btu	210,000 Btu	252,000 Btu
1000 sf/90 sm	105,000 Btu	157,000 Btu	210,000 Btu	263,000 Btu	315,000 Btu

To calculate the true cost of operating your pool, however, don't forget the cost of electricity for pumps, motors, and other appliances. Of course you could also get into the cost of depreciation of the equipment, depletion of chemicals, water use, etc., but then no one would have a pool.

Finally, to measure the exact volume of gas used, you can read the gas utility meter before and after a certain time period when you know the heater is the only gas appliance in operation.

> **TRICKS OF THE TRADE: HEATER GENERAL SIZING CHART**
>
> **Based on the square footage (sf) or square meters (sm) of the pool surface...**
>
> **...and the desired rise in temperature (shown in deg. Fahrenheit or Celcius)...**
>
> **...and using an 8-hour filter cycle for pools:**
>
> **Select a heater with a Btu rating as indicated (or higher)...**
>
> **...and interpolate as needed for values between those shown.**

Repairs and Maintenance

Heater troubleshooting is both art and science. After years in the field, most good service technicians can sense what is wrong by the customer's description or by a look at not just the heater, but the overall pool and the equipment area. Since nothing replaces experience in this field, I will limit my comments to the most common heater failures and how to repair them.

I have learned a lot over the years by watching the technicians that come out on factory service calls. If I couldn't fix a heater, I would call the factory on behalf of the customer and make the appointment at a

---

TRICKS OF THE TRADE: HEATER REPAIR

Some basic guidelines for any heater repair.

- My Golden Rule of heater repair is that you always turn the heater off when making repairs. Preferably turn the pump off as well and disconnect or turn off any source of electricity. If you don't, you might complete the repair or touch some wiring together causing the heater to restart when you don't really want it to. By shutting everything down, you control the entire process—you check your work and control the test-firing when you think you're done. Otherwise the heater controls you.

- It is generally better to replace components rather than repair them and, if it's a well-worn heater with other parts that will soon give out, replace the entire heater instead of adding new parts every month ad infinitum. Associated with this point is the fact that each part in a heater fails for a reason, rarely old age. When you replace or repair something, find out why it failed in the first place or it will happen again. I have been told that I was a genius when I got a heater running again and told what an idiot I was when I had to come back two weeks later for the same repair.

- Most heater repairs are not the heater at all. The majority of heater failures are the result of dirty filters (and add to that low or obstructed water flow from the pool). In short, nothing to do with the heater at all. Moral of the story? Look around at the entire installation before starting on what appears to be a heater problem.

- One other small point that should be obvious but perhaps isn't. If the heater has been running prior to any repair you are making, watch out for hot components. As noted previously, pilots generate over 1100°F (593°C), so they stay hot for a long time after they've gone off. Cabinets and other metal parts get hot too, so watch what you grab.

---

time when I could be there to watch them work. Most of these people really know what they're doing and they have a wealth of knowledge for you to tap.

Basic Start-Up Guide

RATING: EASY

When you first fire up your heater, a few pointers are worth noting.

1. **Bleed** If the gas supply line is new, bleed the air out of the pipe by opening the line at the union near the heater and opening the

shutoff valve. When you smell gas instead of air coming out of the pipe, shut off the valve and reconnect the union. There is still air in the remaining foot or two of pipe and in the combination gas valve inside the heater, but this will bleed quickly if gas is present in the line up to the shutoff valve. Open the gas shutoff valve.

2. **Pump** Make sure the heater on/off switch is off. Turn on the pump and motor and make sure the air is out of the water system. Check for leaks.

3. **Pilot** Light the pilot (refer to the instructions that follow or those provided with the heater) if the unit is a standing pilot type. It might take 60 to 90 seconds to get a flow of gas in the pilot tube as the air bleeds out and is replaced by gas. Never try to light an electronic ignition pilot with a match or other fire source. Believe it or not, I have gone on heater service calls where other service technicians have left the evidence of their stupidity in the heater—burnt matches or burnt paper. After lighting the pilot, turn on the on/off switch and turn up the thermostat as needed to fire the heater.

If it is an electronic ignition unit, turn the valve on the combination gas valve inside the heater to on. Turn on the on/off switch and turn up the thermostat as needed to fire up the heater. It might take a minute or two to bleed the air out of the system and replace it with gas, so don't worry if it seems to take awhile for the pilot to ignite and the unit to fire the first time.

4. **Fire Up** When the heater first fires, the heat will burn off the oil that is applied by the factory to the heat exchanger as a rust preventive. Light smoke for a few minutes is normal. Also normal is some moisture condensation as very cold water runs into the very hot heat

---

TOOLS OF THE TRADE: HEATERS

- Flat-blade screwdriver
- Phillips screwdriver
- Hacksaw
- PVC glue
- PVC primer
- Pipe wrench
- Teflon tape
- Silicone lube
- Needle-nose pliers
- Hammer
- Emery cloth or fine sandpaper
- Pipe wrench
- Multimeter (and millivolt tester if separate)
- Nut driver set
- Electric drill with reaming brush
- Channel lock-type pliers
- Knee pads

exchanger. The condensation will drip down onto the burner tray and sizzle—a little of this is normal too.

5. **Look** Observe the heater for the first 10 minutes. Make sure the smoke and condensation stop and that there are no leaks. Use your eyes and nose (gas leaks?). Turn the heater off and on a few times to be sure it operates properly. Do this from the on/off switch and the thermostat a few times. Remember, if the water is colder than about 65°F (18°C), the thermostat won't turn the unit off because it doesn't register that low. Now turn the pump off. The heater should shut off within 5 seconds. If it doesn't, the pressure switch needs adjustment.

   Use common sense. As the heater is working, look around and make sure you have sufficient ventilation and air supply and that any overhanging building areas or trees are far enough away to stay cool.

6. **Feel** With the pump running, touch the inlet pipe to the heater and then the outlet. The temperature differential should not be more than 10°F (5°C). If it is much more than that, refer to the discussion of bypass valves that follows.

Terry's Troubleshooting Tips

This is the only section of the book I name after myself, because after years of troubleshooting heaters I have developed a method that seems to work for me. I think it will work for you too.

To initiate repairs, you first need to understand how to identify the problem. I suggest you read over the entire repair section to familiarize yourself with component operation and repair; then the troubleshooting tips will make more sense.

While troubleshooting heaters is fairly simple, remember that symptoms are similar for several different problems, so you need to be sure you have isolated the correct malfunction and also be sure there's not more than one malfunction. As mentioned previously, there are so many potential combinations of causes of heater malfunction that you must look at the entire system and installation and be flexible in your thinking.

What follows is by no means comprehensive, but is the way I approach heater problems. This outline will get you through most of the main failures and their causes so that at the least, even if you can't fix the problem yourself, you will be able to communicate intelligently with the factory service technician.

If you do end up calling the factory service office, be prepared to give them a concise description of the problem, the model and serial number of the heater, and the date it went into service if you know. Warranties differ between manufacturers and even within one heater some components are guaranteed for one year, others for five. In the reference sources section at the back of the book you will find the websites of major heater makers for technical help over the phone or factory service (and training seminars) in your area.

I think the simplest way to approach this section is to provide you with my checklist, then to explain any tests or repairs that are not self-explanatory in more detail at the end. This list deals mainly with gas-fueled heaters, because these are the most complex and common in the field. Some of the technology and diagnostic tips apply to any heater, however. So, whatever the problem reported, the steps I follow are in about the same order. Think like a doctor—whatever the patient reports to be wrong, the doctor gives the entire body an exam to see if there are interrelated disorders.

With heaters, there are three basic complaints: the heater won't come on, the heater won't stay on long enough, or the heater won't shut off. Regardless, it pays to check out the entire flow of water, gas, and electricity through the heater because, as mentioned before, one problem might be a symptom of a larger problem. In each area, use your eyes, ears, and nose—look for leaks or soot, listen for odd sounds, and smell for burning components or leaking gas.

Here's how I approach the checkup.

TERRY'S TROUBLESHOOTING CHECKLIST

RATING: ADVANCED

1. Check the water system.
   A. Is the pump primed and running without interruption?
      (1) Is there enough water in the pool?
      (2) Has the air been purged from the system?
   B. Is the water flow too low?
      (1) Are the skimmer and main drain clear?
      (2) Are the pump strainer pot and impeller clear?
      (3) Is the filter clean?
      (4) Are all valves open?
   C. Is the water flow too great?
      (1) Is the internal flow control valve operating?
      (2) Are the in and out water temperatures within 10°F (5°C)?
      (3) Is the external bypass valve set correctly?
   D. Is the water chemistry correct?
      (1) High pH could mean scale in the heater.
      (2) Low pH could be causing leaks.
   E. Are there any visible leaks?
      (1) At the exterior plumbing connections?
      (2) At the interior heat exchanger components?
2. Check the gas system.
   A. Is gas getting to the heater?
      (1) Look for the pilot flame.
      (2) Conduct the open gas line test at the heater.
      (3) Is the propane tank full?
      (4) Are the gas supply lines adequate and unobstructed?
      (5) Is the supply dedicated (only used for the heater)?
      (6) Is the heater large enough for demand and expectation?
      (7) Is the combination gas valve turned to on?
   B. If the pilot and/or burner are working. . .
      (1) Is the flame 2 to 4 inches (5 to 10 centimeters) and steady?
      (2) Is the flame a healthy blue?
      (3) Are all the burners fully lit?
      (4) Is pilot ignition within 5 seconds?
      (5) Is burner ignition within 10 seconds after pilot?
      (6) Is the tray ignition without flash or a loud boom?

C. Is the ventilation adequate?

    (1) Is there adequate air supply to the heater?

    (2) Is there adequate venting away from the heater?

D. Smell for leaks.

    (1) Sniff around the outside connections and unions.

    (2) Sniff around the combination gas valve and joints.

3. Check the electrical system (Figs. 6-9 and 6-10).

  A. Is the on/off switch on? Is the remote switch (if any) on?

  B. Is the thermostat turned up high enough?

  C. Is 25 volts coming out of the transformer?

    (1) Is 120/240 volts coming into the transformer?

    (2) Is the heater grounded properly?

    (3) Are all the connections tight and clean?

    (4) Is the reset breaker button tripped (electric-fueled heaters)?

  D. Check the pilot.

    (1) Light the standing pilot and look for a healthy flame; or

    (2) Electronic ignition has fired the pilot and there is a healthy flame.

    (3) Is the pilot clear of rust, dampness, or insects?

  E. Check the control circuit. Follow the path of the electricity.

    (1) Is there power from the pilot generator or transformer?

    (2) Is there power to each switch. . .

        (A) On/off?

        (B) Thermostat?

        (C) Fusible link?

        (D) Pressure switch?

        (E) High-limit (two)?

        (F) In-line fuse (some models)?

        (G) Fireman's switch?

  F. Is there power to the gas valve?

The troubleshooting checklist can also serve as a good preventative review for any heater. Once a year it's a good idea to methodically check out the health of the heater by running through the entire list.

But what if specific symptoms of a problem occur and you don't have time to perform a thorough "physical" on the heater? Here are the

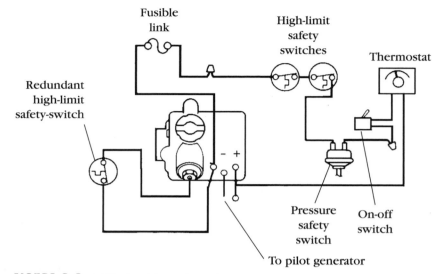

FIGURE 6-9 Millivolt wiring schematic. *Image courtesy of Jandy.*

four most common problems and probable solutions, listed in order from the most common to the most obscure. Detailed descriptions of these repair procedures are found in *The Ultimate Pool Maintenance Manual*. Remember, symptoms often warn of more than one item that needs attention:

Pool not reaching desired temperature

- Set the thermostat higher.
- Run the circulation system longer and/or set the time clock so that the water heats up before the most common bathing time.
- Clean the filter or remove other circulation obstructions.
- The heater may be undersized (see the section on selection).
- If burner flame is low ("lazy"), check gas pressure or clear debris from burners.

Heater makes whining/knocking noise and/or stays on after pump goes off

- Pressure switch out of adjustment.
- Clean heat exchanger.

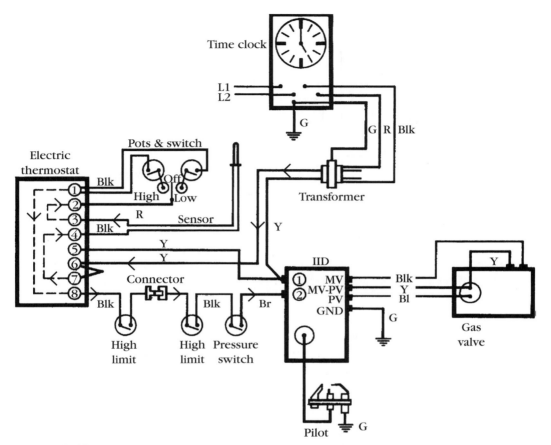

FIGURE 6-10 **Electronic ignition heater control circuit (with electronic thermostat).** *Raypak, Inc.*

Heater goes on/off frequently

- Adjust pressure switch.
- Check circulation, clean filter.
- Adjust bypass valve (or check that automatic bypass is undamaged).

Soot forming inside heater

- Air supply restricted—clear.
- Flow rate excessive—correct.
- Clear debris from burners.

TRICKS OF THE TRADE: LIGHTING THE PILOT

This procedure applies to standing pilot units only, because the electronic ignition is automatic unless something is broken. Instructions are almost always printed on the heater itself. Look for these and follow them. If the directions for your particular heater are obscured or missing, the following procedure is most common:

1. **Shut off:** Turn the gas valve control to Off and wait 5 minutes for the gas in the burner tray or around the pilot to dissipate (safety consideration). Turn the on/off switch to Off.

2. **Listen:** Turn the gas valve control to Pilot and depress. If the area is quiet, you should hear a strong hissing sound as the gas escapes from the end of the pilot. If not, it might be clogged by rust or insects.

3. **Light:** Light the pilot and continue to depress the control.

4. **Wait:** Hold the control down for at least 60 seconds. This allows heat from the pilot to generate electricity in the thermocouple to power the gas valve (which will then electronically hold the pilot gas valve open and power the control circuit which ultimately opens the main burner gas valve). Release the control and the pilot should remain lit.

5. **Verify:** This step is not usually included in instructions printed on the heater—that the pilot flame is healthy; that is, look to see that a strong-burning blue flame of 2 to 3 inches (5 to 7 centimeters) extends toward the burner tray and a secondary flame of equal value is heating the thermocouple. Make sure the pilot is securely in place; it might be burning properly, but rusted fasteners or poor installation from a previous repair might have left it dangling near the burner tray, but not in close contact with it. Burner tray design leaves a lot to be desired when servicing the pilot assembly of most heaters. A common problem is that a lazy service technician previously left the pilot assembly hanging in the general area by wire, tape, a screw not fully in place, etc. The result is that the pilot burns just fine, but the burner tray needs to be flooded with excessive amounts of gas before any reaches the pilot flame to be ignited. The resulting ignition is explosive. Therefore, always make this visual inspection.

6. **Fire up:** Turn the gas valve control to On. Stand back from the heater (and to one side—not in front of the open front of the heater) in case of flashback (the explosive ignition described previously). Make sure the pump is operating and water is flowing freely, turn the on/off switch to On, and turn the thermostat up. The heater should fire normally. This pilot lighting procedure does not normally apply to millivolt heaters with piezo ignition units. Piezo units create a spark near the pilot light to ignite the gas, simply by turning on the gas supply as previously described and then pushing the piezo's button. You have probably used this system on a gas barbecue and have probably also noted how often piezo units become fouled with dirt, cobwebs, or rust and fail to operate. In these instances, you will need to light the pilot manually as just described, providing a spark for the pilot gas supply by a good old-fashioned match.

FAQS: HEATERS

Do I need both a heater and a cover?

• No, but a cover and a solar heater will insulate and warm your pool significantly, so a gas heater doesn't need to work as hard (or use as much fuel, which could save you some money). The cover also keeps the pool clean and reduces evaporation, which is exacerbated by heating in the first place. In order of value and importance, you should consider a cover, solar heating panels, and finally a gas heater.

Will a pool heater work on electricity, propane, or natural gas?

• Yes, but an electrically fueled heater is slow to heat the pool and the fuel costs are considerably more than for natural gas or propane. Natural gas or propane will heat the pool about the same, but the heater needs slightly different burner and gas regulating components, so be sure to use the heater that matches the fuel in your area.

How long will it take to heat my pool?

• You will feel warm water coming from the return outlet immediately when heating with solar or gas heaters. Depending on the size of the pool and the capacity of the heater, you should be able to take the chill off the pool and make it swimmable within a day, especially if insulated by a cover.

How much does it cost to run a heater?

• Depending on the size of your pool and how warm you want the water to be, you can heat your pool for less than $5 per day in the coolest months of spring to make the water comfortable for swimming. As the weather warms and especially if you have a cover, that cost will drop significantly.

Preventative Maintenance

RATING: EASY

The best preventive maintenance is to use the heater regularly. As noted previously, corrosion, insects, nesting rodents, and wind-blown dirt create many heater problems that can be eliminated by regular use. The heat helps to dry any airborne moisture that might otherwise rust the components. It discourages insects and rodents before they get too comfortable. It keeps electricity flowing through the circuits, preventing corrosion that creates resistance that might ultimately break the circuit completely. It burns off the odd leaf or debris that lands inside the top vents before a greater accumulation builds up, which when the heater comes on might start a fire or send flying embers into the air. In short, running at least a few minutes a day is great therapy for a heater.

Otherwise heaters need only be visually inspected from time to time. A look around will detect sooting, gas or water leaks, or other problems before they begin. Keep leaves and debris off the top of the heater. Look at the pilot and burner flames. Are they strong, blue, and burning straight up at least 2 to 4 inches (5 to 10 centimeters)? Open the drain plug on the heat exchanger and look for scale buildup.

A Reminder

Always remember that a heater is potentially the most dangerous component of pool and spa equipment. Whenever you make a repair, consider **safety first**. Refer back to the safety checklist in this chapter before you start work on a heater.

Other Equipment and Options

We have now examined the major components of water circulation, filtration, and heating. There are several pieces of equipment that complete the mechanical resources available to the pool.

Time Clocks

A key component of any water maintenance system is a time clock. Time clocks make sure that water is circulated through the filter each day, that it is heated sufficiently and/or at the correct time of day, that decorative lights come on at the right time, that automatic cleaners operate daily, and a host of other useful functions.

Electromechanical Timers

Figure 7-1 shows an exploded view of a typical 120-volt time clock, and Fig. 7-2 (left) shows the same product assembled. It allows various on or off settings throughout each 24-hour period. Some models also include a seven-day feature, allowing you to determine which days of the week you want the clock to control the system. The only difference between a 110-volt and a 240-volt clock is that the 240-volt unit does not have a neutral line, but includes two incoming lines and two outgoing loads for the equipment. The neutral location is a ground. For explanation purposes however, this diagram shows the basic parts found in any time clock.

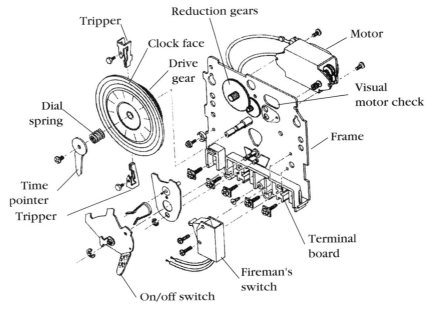

FIGURE 7-1 Typical electromechanical time clock. *Intermatic, Inc.*

FIGURE 7-2 The electromechanical time clock (left) and twist timer (right). *Intermatic, Inc.*

The small electric motor runs on the same push-pull concept of opposing magnets as the motors used for pumps. A winding is electrified that creates an electromagnetic field and sets up the rotation, which happens at a predictable 400 rpm.

The 24-hour dial would not operate very well at that speed, so a series of reduction gears in the motor housing result in the actual drive gear turning at a speed that rotates the dial just once every 24 hours. In this way, you can tell the time of day by the clock face and use it to set on or off switching of the system.

Most ordinary watches or clocks have a face that remains fixed, while the hands of the clock rotate to give the time. The time clock operates on the opposite concept—it has only one hand, which remains stable, and a face that rotates.

Note the on and off trippers (the words *on* and *off* are etched on the trippers). By setting these where you want the system to go on or off, the clock controls the flow of electricity to your system accordingly. More than one set of trippers (also called *dogs*) can be affixed to the dial, so the system can go on and off several times during a 24-hour period. As you can see on the top tripper, the nipple on the bottom of the tripper is located almost midway along the edge, while the nipple on the lower tripper is located at the end. These nipples engage the cam on the on/off switch of the clock, turning it on or off as the clock face rotates.

To set the clock, you simply pull the face toward you and rotate it until the number on the dial, corresponding to the correct time of day, is under the time pointer. By pulling it forward, its drive gear is disengaged from the motor drive gear and it rotates freely. As noted earlier, the time pointer is the only hand on the clock. The dial is divided into each of the 24 hours of the day (differentiating between a.m. and p.m., some clocks say Day and Night) and each hour is further divided into quarter hours, so when setting the correct time of day you can be fairly precise. After setting the correct time, release the dial and it snaps back into position, reengaging its drive gear with the motor drive gear.

When the switch lever moves to the left, it lifts the contacts apart, breaking the electrical circuit. When the lever moves back to the right, the contact arm is allowed to drop back into contact with the lower contact, completing the electrical circuit. Simple, eh? The lever also allows manual operation, so regardless of where the trippers are set you can manually operate the system.

The contacts are attached to screw terminals for attaching the wires of the appliance to be controlled. On a 120-volt model, the hot 120-volt wire is attached to the terminal marked Line and the neutral to the terminal marked Neutral. The hot wire for the load (the appliance) is connected to the load terminal and its neutral is connected to the neutral terminal. Some clocks provide separate neutral terminals for the line and load, but they both join to the same wire that returns to the circuit breaker. As noted, the 240-volt version includes two line terminals for the two hot lines coming in, and two load terminals for the supply to the appliance.

The wires feeding electricity to the clock are wrapped on the line and neutral terminals (with a 240-volt clock, the one lead would be attached to each line terminal). This way there is a constant supply of electricity to the clock.

Waterproof boxes in metal or plastic are available to house time clocks. The unit shown in Figs. 7-1 and 7-2 simply snaps into clips built into the box. Other versions have built-in brackets that are designed to align with the screw holes on the clock plate, and these are held in place by two machine screws. The boxes are built with knock-out holes to accommodate wiring conduit of various sizes and have predrilled holes in the back for mounting the box to a wall.

Intermatic makes most of the pool and spa time clocks you will encounter. Paragon is another maker, and Len Gordon makes a miniature version included in its clock and control packages. As with other components of pool equipment, if you understand the parts and operation of one, the others are easy to figure out.

Twist Timers

Twist timers (Fig. 7-2, right) are used mostly with lights for a limited amount of operation as the user demands. A twist

---

TRICKS OF THE TRADE: TIME CLOCKS

- A quick way to tell if power is getting to the clock (assuming the clock motor is in working condition) is to look through the opening provided marked Visual Inspection, or words to that effect. Look in here to see the motor gears spinning. Some clocks have the motor mounted on the front of the clock and the center hub of the drive gear is visible. Close inspection will reveal if the hub is spinning or not.

- When a second clock is used to run an automatic pool cleaner booster pump, make sure it is set to come on at least one hour after the circulation pump comes on and is set to go off at least one hour before the circulation pump goes off. These boosters rely on the circulation pump for water supply, and without it will burn out their components. When checking or resetting the system clock, be sure to follow this guideline with the booster pump clock.

timer is built to fit in a typical light switch box and contains no user serviceable parts. This unit has a faceplate showing 15, 30, 45, or 60 minutes. The knob attached to the shaft that comes through the faceplate is twisted until its pointer or arrow aligns with the desired number of minutes. The circuit is completed in any position except off and the mechanical timer is spring-loaded to unwind for the number of minutes selected. When the spring is unwound, the circuit is broken and the appliance shuts off.

Twist timers are available in 120 and 240 volts and are used in place of a simple on/off switch, usually where users might forget to turn off such a switch. Another good use for a twist timer is for yard lights, where the user must set the desired time so that the lights will go off even if the user forgets.

In the repairs section later, I don't discuss twist timers because, as noted, there are no user serviceable parts. If they fail, replace them. It is no more difficult than replacing a light switch, and all you need is a screwdriver. When you buy the replacement, follow the instructions in the box if you don't find it self-explanatory. As with any electrical repair, be sure the electricity is disconnected at the breaker panel.

Electronic Timers

Do you own a VCR? Are you one of those people who never sets the time of day on it, but lets it flash 12:00 all the time? Well, when you learn about electronic timers for water system controls, you will also know how to set your VCR.

Electronic timers are used in some pool control packages that also include electronic thermostats and other sophisticated controls.

Repairs

There's not much that goes wrong with electromechanical time clocks and it's easy to service most of them.

REPLACEMENT

RATING: EASY

When a time clock rusts or the gears wear out and the unit needs to be replaced, it

---

TOOLS OF THE TRADE: TIME CLOCKS AND GFIs

- Screwdriver set
- Loose-joint pliers
- Needle-nose pliers
- WD-40 spray lubricant
- Multimeter electrical circuit tester

takes longer to buy the new unit than to perform the replacement. When buying the replacement, be sure to get the same voltage clock as you have in the existing installation. Also, buy the same make of clock, because a different manufacturer's clock probably won't fit in the existing box.

1. **Power** Turn off the power supply at the breaker.

2. **Disconnect** Remove the line and load wires from the old clock and any ground wire. Remember (or mark) which wire is which.

3. **Replace** Unscrew the holding screws or unclip the old clock to remove it from the box. Snap or screw the new one in place.

4. **Connect** Reconnect the wires to the appropriate line or load terminals. Replacement clocks might not be arranged the same as the old one, so be sure you are getting the line wires (two hots or one hot and one neutral) attached to the line terminals and the two appliance wires attached to the load terminals. If it is not clearly marked, the simple way to tell which terminal is which is to notice where the clock motor is connected—the two clock wires are always attached to the line terminals.

5. **Test** Turn the power back on and test the clock operation by turning the manual on/off lever to On.

Most manufacturers sell just the clock motor for replacement. I have found that if the motor is old enough to burn out, the clock mechanics are probably wearing out too, so it pays to replace the entire mechanism. Besides, the $20 or so price difference between the motor and the entire clock unit is not worth messing with motor replacement only. In fact, it takes longer to replace the motor only than it does the entire clock.

By the way, if the power supply goes to an on/off wall switch or other device first, the power might be interrupted, meaning the clock cannot keep accurate time. Keep in mind that the clock motor itself must always have a source of power.

CLEANING

RATING: EASY

Sometimes insects will nest in your time clock. I have found countless times when a clock fails to turn a system on that it is clogged with nesting ants. The cure is simple: turn off the power and remove and clean the contacts. When finished, apply a liberal amount of insecticide to the interior of the housing before reinstalling the mechanism. Don't spray the mechanism itself. You might create unintended electrical contact through the liquid.

Other than that, corrosion will occasionally bind up a clock. If you know it is getting power but cannot see the gears turning, take the tip of your screwdriver and gently force the gear (the one visible through the inspection hole) in a clockwise direction. Often that will get it started and it will continue to run fine after that. If it happens once, it will probably happen again unless you give the gears a good general lubrication.

To lubricate a time clock, turn off the power source and remove the clock from the box (wires still attached) to expose the rear of the clock. Spray WD-40 liberally around the gears. Do the same on the front to lube the gears behind the dial face. As noted, be careful not to wet the electrical contacts. Put the clock back in the box and turn the power back on. Turn the clock on and off a few times to work in the oil. If the clock fails again, replace it.

MECHANICAL FAILURES

RATING: EASY

When you pull out the dial face of the clock to set the time, take care that when you release it you get a true reengagement of the dial with the motor drive gear. Try setting different times on the clock and you will note that sometimes the two gears don't mesh, but rather the dial gear sits on top of the motor gear. Obviously, in this case the clock won't work.

The answer is to wiggle the dial face as you release it. As you release it, twist the dial back and forth very slightly in your hand to make sure the gears mesh. With a few practice settings you'll feel the difference between a dial that has gone back into place completely and one that is slightly hung up. Often, clocks that don't work are a result of this setting problem, so make this one of the first things you check when you suspect time clock failure.

The second mechanical problem of time clocks is with the trippers. If the screw is not twisted tightly on the face of the clock, they come loose and rotate around the dial, pushed by the control lever instead of doing the pushing themselves. Check the trippers regularly because they can come loose over time or from system vibration.

Also, if you are trying to set a time that happens to be close to a tripper setting, the dial will not engage with the motor gears. You need to move the tripper a little to make room for the dial gear. Finally, trippers can wear out, so if the clock is keeping good time but not turning the system on or off, try a new pair of trippers.

SETTINGS

RATING: EASY

When setting on/off trippers, you can place them on the dial face side by side to what appears to be about 30 minutes between them. I have found that when they are too close, they won't operate the lever. Generally, trippers must be at least an hour apart to operate.

Make frequent checks of your time clocks. Power outages, someone working at the house who shuts off all the power, the twice yearly daylight savings time changes, and any number of other household situations can interrupt power to the time clock. Each time this happens, the clock stops and needs to be reset when the power returns in order to reflect the correct time.

Ground Fault Circuit Interrupters

The Ultimate Pool Maintenance Manual devotes an entire chapter to electrical theory, components, and circuit testing. Since aboveground pools typically deal with only household current via one cord feeding a skidpack, that level of detail is unnecessary here, but if you have more elaborate equipment, refer to that book. The one component of electrical circuitry that needs describing for aboveground pool owners is the one that deals with your personal safety—the ground fault circuit interrupter, or *GFI* for short.

When equipment or wiring fails it might draw more current than the appliance can use, burning out the appliance. The circuit breaker is designed to break the circuit when demand exceeds the rating of the breaker. It takes so little current to kill a human that the typical breaker will deliver a lethal dose before breaking the circuit. In other words, circuit breakers are designed to protect equipment, not humans.

The GFI is designed to protect humans. It is a circuit breaker that detects problems at a low enough level to protect you before lethal doses are delivered. It breaks a circuit when it detects a ground fault. The GFI constantly measures the current going out of it (to the appliance) and coming back into it. If an inadvertent grounding takes place, such as if the metal case of an appliance were electrified, and you touched it, completing a pathway for current to the ground, the GFI would detect the drop in the current it was receiving and break the circuit. The GFI detects variations as low as 0.005 amp, which is about half the lethal charge to a child and about one-sixth a lethal dose to you. The GFI cuts the circuit within one-fortieth of one second, so it is not only sensitive, it's quick.

There are three basic styles of GFI designed for use in any one of three points in the circuit. The first looks like a standard circuit breaker in the electrical panel, but it has a test button in the face of the breaker in addition to the on/off breaker switch. By pressing the test button, you are simulating an unbalanced current condition inside the breaker and thereby testing the efficiency of the GFI. The GFI breaker resets the same way a normal panel breaker does.

The second type of GFI is built into a wall outlet, which is the type you might install for plugging in your skidpack (refer back to Fig. 5-13). It also contains a test button and a switch to reset the GFI.

The third type is a portable GFI, a unit that plugs into a wall outlet. The appliance is then plugged into the GFI, making the outlet a GFI outlet.

All types of GFIs, like any other mechanical device, are subject to failure and should be tested at least every month.

If they provide so much safety, why aren't all breakers and outlets GFIs? The first answer is probably cost, since the GFI breaker or outlet costs two to four times as much as a standard one. The practical reason is that some appliances or circuits operate normally with slight variations in current, so the GFI would constantly be breaking circuits for the wrong reason.

In fact, slight variations might occur in the pool equipment, causing the GFI to trip even though everything is functioning properly. Panel GFIs are somewhat prone to this problem, because often there is slight current leakage inside the panel, and long wire runs to the equipment location can create slight variations in current that cause the GFI to trip. For this reason, it is best to locate the GFI as close to the appliance(s) as possible.

The National Electric Code (NEC) specifies that electrical outlets located within 15 feet (4.6 meters) of the water's edge must be protected by a GFI and that circuits for all underwater lighting be so equipped. Underwriter's Laboratory (UL) requires all portable pools and spas be wired with a GFI.

Ladders, Slides, Safety Barriers, and Decks

A great way to add fun and safety to your above-ground pool is to add an entry system, including ladders and slides, some fencing (called *safety barriers*), or a deck.

Ladders

Ladders and handrails are a necessity for above-ground pools and are often included as part of a package when you purchase the pool. Others make excellent optional improvements to the safety and utility of your investment.

TYPES

Ladders are typically A-frame units that straddle both sides of the pool wall to facilitate both entering and exiting the pool. They are made of heavy-duty plastic (Fig. 7-3A) or metal with plastic step covers (Fig. 7-3B) to prevent burning your feet. Both styles incorporate a handrail and a platform at the top of the ladder, but remember that the platform is not meant for jumping or diving.

Figure 7-3C shows a more elaborate ladder with a fence and gate built in to limit access to the pool. This model is installed against the side of the pool, and you can install another step unit in the water to facilitate entering and exiting the pool itself. Figure 7-3D shows a popular fiberglass "wedding cake" model in the pool.

Ladders are also commonly incorporated into a prefabricated deck unit (refer

---

TOOLS OF THE TRADE: LADDERS, SLIDES, SAFETY BARRIERS, AND DECKS

- Screwdriver set
- Box wrench set
- Socket wrench set
- Channel-lock type pliers
- Carpentry tools (as needed for specialty wooden decks)

FIGURE 7-3 (A) Typical plastic ladder. (B) Typical metal ladder. (C) Gated ladder. (D) Full entry step and rail system. *C: Delair Group, LLC. D: Hoffinger Industries/Doughboy Pools.*

back to Figs. 1-4, 1-14, 1-16, and 1-23G). For soft-sided/framed pools, the ladder is designed to hang from the framework or be set up in an A-frame style as shown in Fig. 1-9.

INSTALLATION AND MAINTENANCE

RATING: EASY

Because there are so many makes and models of ladders for above-ground pools, there is no typical assembly or installation guide. That said, most ladders are snap-together versions with very little hardware, requiring no skill and only a few minutes to assemble.

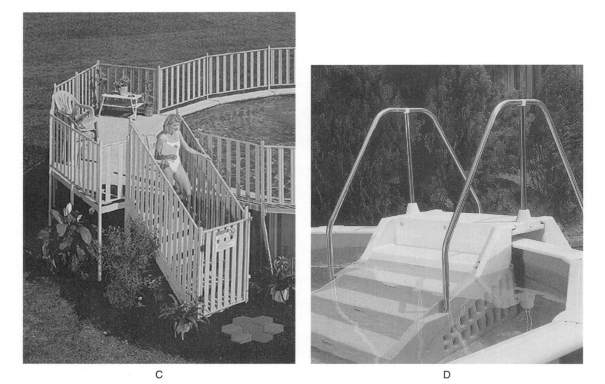

C   D

FIGURE 7-3 *(Continued)*

Installation is a matter of following the manufacturer's guidelines, but the one element all ladder units have in common is that they need to be mounted on firm, level ground. Uneven ground, which may get wet from splashing when the pool is in use, will result in collapses and injury. The other key to successful installation is to use all warning stickers and follow the advice they give. Don't allow ladders to be used for anything except entering and exiting the pool, one swimmer at a time. Overloading a ladder or allowing rough play on the steps by several children will cause the unit to fail over time.

Maintenance is equally a common-sense matter. Keep the plastic clean using mild soap and water (keep the soap out of the pool!), but the most important aspect of maintaining a ladder is inspection. Every week, make a careful inspection of the components, especially where they join to one another. Tighten any hardware that has come loose and look carefully at both sides of the ladder because the steps may wear unevenly over time. Look for signs of rust on metal components. Even aluminum can decay over time, so tap on metal components to ensure that they are still sound. Don't tap with sharp tools because you may scratch the finish that is designed to prevent rust in the first place. If you see any significant loosening or wearing of components, especially plastic ones, discontinue use until you can replace the defective component or the entire ladder unit.

Slides

Slides can be a great addition to a family pool, but they require close supervision and common-sense maintenance to ensure that they are used safely. BE SURE YOUR POOL IS DESIGNED FOR USE WITH A SLIDE. Please heed this safety warning, and be sure that your pool manufacturer has recommended a slide and that the slide you choose is designed for use with your particular pool. Serious injury can occur, and if you damage the pool, it can void the warranty.

Slides are generally made of fiberglass with metal frames and steps (Fig. 7-4). A straight slide will be 8 to 13 feet (2.5 to 4 meters) long, requiring a lot of deck space. If deck or yard space is limited, left-handed or right-handed curved slides are available.

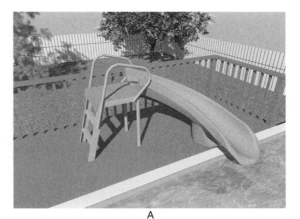

A

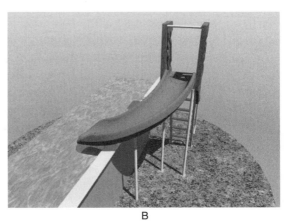

B

FIGURE 7-4 Slides: (A) deck-mounted; (B) ground-mounted. *Summit-USA, Inc.*

INSTALLATION

RATING: EASY

Slide installation is simple, and anchor kits are available at your supply house with instructions. Following the instructions in the kit, anyone can install a slide.

Some slides are plumbed to provide a sheet of water along the slide area to make them more slippery. These can be plumbed into a water supply line, but prolonged use will lead to an overflowing pool. A better installation is to plumb the slide to a return line from the pool's circulation system. This might require running a long and complicated pipe from the equipment area to the slide, but the supply pipe is usually ¼-inch (6-millimeter) diameter, so concealing it in gardens or along building edges is not difficult. Always fit the line with a shutoff valve so the slide can be turned off when it is not in use. The water running over the large surface area of the slide, which is usually quite hot, will evaporate rapidly, slowly draining the pool.

MAINTENANCE
RATING: EASY

Because of constant exposure to the elements, slides often discolor. Your supply house sells a glaze/polish kit that can restore the appearance to almost new. I don't recommend painting slides, even with epoxy paints, because the constant sliding friction and exposure to sun and temperature extremes will quickly make the painted surface look worse than the original problem. Painted surfaces often oxidize and your swimmers will end up with powdery streaks over the backs and bottoms of their swimsuits.

Finally, check the water supply line (if the system has one) for leaks that might deplete the pool water level, leading to equipment that runs dry and expensive repairs.

Safety Barriers

Many jurisdictions now require some type of fence, solid cover, or other safety barrier around a pool to prevent drowning ("barrier codes"). If more people paid attention to this common-sense requirement, many needless deaths and injuries could be prevented every year.

Generally, barrier codes are written for "swimming pools" (including above-ground pools) and include at least these requirements:

- The pool must have a fence on all four sides at least 4 feet (1.2 meters) high.

- Gates must be self-closing and latching.

- Where the home itself forms one or more "walls" of the fence, doors must have locks or other provisions to prevent children from accessing the pool.

■ Sometimes an alarm can be substituted for certain barrier requirements (check on the type of alarm permitted in your area).

Remember, many drownings of children occur even when barriers are provided if gates are left open or covers left off. Safety barriers are only as good as their actual use.

Many deck or ladder units incorporate the safety barriers right into their design. Figure 7-5 shows a removable safety barrier that fits easily into special brackets designed to be bolted to the pool's uprights. These are typically metal and can be installed or removed by one person in a few minutes (after the brackets have been installed on the uprights, which is done only upon the initial installation). Other safety barriers are integrated into deck structures as shown in the next section.

FIGURE 7-5 **Snap-on safety barriers.** *Hoffinger Industries/Doughboy Pools.*

There are no installation or maintenance challenges with simple safety barriers other than the recommendations made elsewhere in this section concerning cleaning and frequent inspection for component integrity.

Decks

A deck is an excellent addition to an above-ground pool because it allows you to keep an eye on children in the water and respond more quickly if there's a problem. It also provides greater enjoyment for nonswimmers, who are part of the pool experience in a way that isn't possible when seated on the ground next to the pool and often below eye level.

Most above-ground pool makers sell deck units that are specifically designed for use with their pools. Throughout this book, pools are shown with specially designed decks attached, including Figs. 1-4, 1-14, 1-16, and 1-23G.

Wood decks are also a popular, if somewhat more expensive, addition to your above-ground pool. Figure 7-6 shows two examples of how wood decks provide beauty, safety, and added value to your yard and pool investment. Since wood decks are all custom jobs, discuss painting, maintenance, and life-expectancy with your carpenter.

Toss Rings

A toss ring (Fig. 7-7), like a life ring on a ship, is a foam plastic ring (17 to 24 inches or 40 to 60 centimeters in diameter) that has a rope attached. It should be mounted prominently near the pool for rescue purposes if someone is drowning. Most jurisdictions require the toss ring for commercial pools only, but I strongly urge you to keep one at your residential pool as well.

Thermometers

I recommend a unit that is built into the skimmer cover (Fig. 7-8) and takes the temperature at the skimmer. You can also simply tie a thermometer to a rail or ladder, or you can use a tube model with a float on the top that floats on the surface of the water. These usually float into the skimmer basket, which is a good place for them to keep them out of sight.

FIGURE 7-6 (A) Typical wood deck and stairs for above-ground pools. (B) Wooden pool deck and fencing. *B: Splash SuperPools, LLC.*

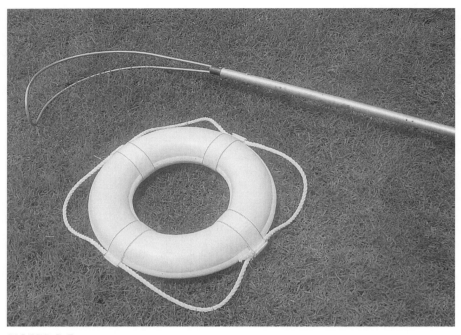

FIGURE 7-7 Toss ring and rescue hook.

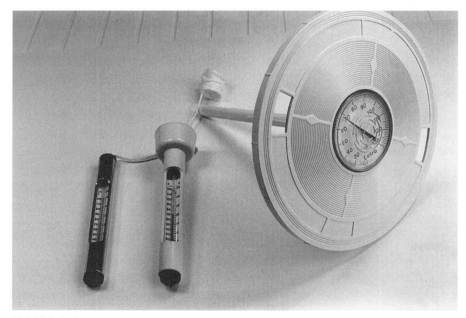

FIGURE 7-8 Styles of thermometers.

In-line thermometers, designed like in-line flow meters, are also available for installation into the equipment plumbing or directly to the heater manifold. For you high-tech techies, digital readout, battery-operated thermometers, and pH testers are also available in floating models or with test probes that you put in the water while reading the information on a small hand-held device about the size of a calculator (see Chap. 8).

Automatic Pool Cleaners

There are numerous designs of automatic pool cleaners available today, but the simple suction-side version is the most common in above-ground pools (Fig. 7-9).

There are two categories of automatic pool cleaner in common use and two other less effective technologies you might encounter. Let's dispense with those two first.

FIGURE 7-9 Typical above-ground pool automatic cleaner.

Electric Robot

The electric robot type of automatic pool cleaner sounds futuristic, but is nothing more than a battery-powered (some are 120 volt, some transformed down to 12 volt) vacuum cleaner with a bag that catches debris as the unit patrols the pool bottom. As you might guess, these are expensive and introduce potentially lethal electricity into the water. Because of the potential for entanglement and injury, you should not use the pool when *any* automatic cleaner is in use, but that precaution is especially important when using an electrically powered unit because it introduces the added hazard of electrocution from a frayed wire if you touch the water at all.

Booster Pump Systems

Booster pump systems take water after the filter and heater, which is already on its way back to the pool, turbocharge or pressurize it by running it through a separate pump and motor, then send this high-pressure water stream through flexible hoses into a cleaner that patrols the pool bottom.

In one style, called a *vacuum head* type, the cleaning device has its own catch bag for collecting debris, much like a vacuum cleaner. In another variation, the *sweep head* type floats on top of the water with long flexible arms that swirl along the walls and bottom, stirring up the debris. A special basket is fitted over the main drain so that the stirred-up debris is caught in either the main drain or the skimmer and any fine dirt is filtered out normally. Because both booster pump variations require an additional pump, plumbing, and electrical connections, they are not commonly used in above-ground pools. The sweep head model is obsolete and is found only on older installations, but one or two vacuum head models are made specifically for above-ground pools. Since these automatic cleaners are rarely used for above-ground pools, we won't delve further into the subject here, but they do work well, and complete instructions for installation, repair, and maintenance can be found in *The Ultimate Pool Maintenance Manual*.

Now that we have reviewed the two less common automatic cleaners, let's take a look at the two models more likely to be in your above-ground pool.

TRICKS OF THE TRADE: AUTOMATIC POOL CLEANERS

- Don't use your automatic cleaner more than necessary. Any mechanical object running around your vinyl liner will cause wear, especially in the patterns of printed liners.

- Strong winds can interfere with proper operation of any automatic pool cleaner because they force hoses and surface float components to one end of the pool. Try again when the wind calms down.

- Always disconnect your automatic cleaner before backwashing the filter; otherwise, suction-side units will send dirt to the interior of filter grids. Boosterless water pressure units may become clogged internally by any debris from the filter when you switch it back from backwash mode to normal circulation.

- Don't allow swimming when any automatic pool cleaner is in use. You can easily become entangled and injured, so this is a safety issue. Swimming also stirs up the dirt and debris your auto cleaner is trying to collect, defeating the purpose of having it in the first place.

- Before putting the catch bag in place, slip the end of an old stocking over the opening and secure it with a rubber band. Then install the catch bag. The fine mesh of the stocking will capture fine dirt and sand, saving you from having to vacuum the pool so often. Some supply houses sell small stockings made for this purpose, which might be easier than constantly raiding a nearby panty drawer.

- Empty the catch bag and/or sock as needed, but don't be obsessive about it. The bag actually works better with some leaves or dirt in it because the debris helps filter as well, catching fine particles in the leaves. Make sure the openings on the bottom and through the center of the unit are not clogged with large leaves so there is always a clear path for the debris to get into the bag.

- Watch the unit perform for a few minutes each time you service the pool. If it runs listlessly or fails to run in a pattern that will clean the entire pool, follow the adjustment procedures outlined in the installation section of the owner's manual.

Boosterless Water Pressure Systems

A variation on the booster concept is a unit that uses the circulation pump itself as the booster, by connecting to the return line as it discharges into the pool. This style is called *boosterless* because it uses no separate water pressure boosting device.

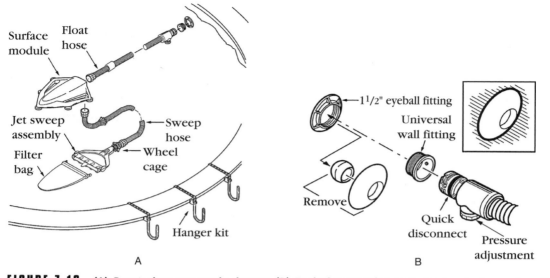

FIGURE 7-10 (A) Boosterless automatic cleaner. (B) Typical connection parts for boosterless automatic cleaner. © *Polaris Pool Systems, Inc.*

One of the leading manufacturers of automatic pool cleaners for both inground and above-ground pools is Polaris. Their boosterless water pressure system is aptly named the Polaris TurboTurtle (Fig. 7-10A), and it can vacuum the bottom and sides of most above-ground pools in 2 to 3 hours. The boosterless water pressure system is superior to the suction-side system (discussed in the next section) because it allows the skimmer to operate when the automatic cleaner is also working.

These units are also popular for their ease of installation and maintenance. Figure 7-10B shows the universal wall fitting, which threads into the return line plumbing of your pool. One end of the automatic cleaner's hose attaches to this fitting with a quick disconnect, and the other end attaches to the surface module, which floats on the surface of the pool. The sweep unit, which actually vacuums the pool bottom and sides, then attaches to the surface module, and you are ready to use your automatic cleaner. Hoses typically come in sections (which can be purchased at your pool supply store if you need replacements or extensions) so you can ensure coverage of the entire pool.

QUICK START GUIDE: INSTALLING A NEW AUTOMATIC POOL CLEANER (SUCTION-SIDE OR BOOSTERLESS MODEL)

1. **Prep**
   - Unpack automatic cleaner and lay hoses in sun to uncoil. Read the owner's manual.
   - Connect surface unit and vacuum unit with hose (boosterless model only), adjusting length of hose as needed for depth of pool plus 50 percent.
   - Shut off pump (tape over switch or breaker so no one can turn it on before you finish).

2. **Set Up**
   - Suction side: Connect suction hose to skimmer suction port.
   - Boosterless: Connect feeder hose to pool return outlet.
   - Attach auto cleaner to the other end of the hose.
   - Walk unit to the farthest part of the pool, and eliminate any excess hose.
   - Place unit in pool.

3. **Start Up**
   - Turn pump back on, and observe auto cleaner.
   - Adjust for optimum performance per owner's manual.

The quick disconnect is fitted with an adjustable pressure relief valve, which is designed to ensure that the unit moves around the pool at a speed that will optimize performance. If the sweep unit doesn't stay on the bottom of the pool, unscrew the valve to relieve excess pressure. If the unit appears sluggish, tighten the valve to increase the amount of pressure in the system. It's really that simple to get excellent automatic cleaning results. If the unit remains sluggish even with 100 percent of the return water flowing through it, you may have a dirty filter or other obstruction in the pool circulation system. There is also a particle screen built into the quick disconnect to trap any fine debris that may escape your pool filter. Check and clean this screen each time you reconnect the hose to the return line.

The debris bag should be emptied regularly for best performance, and remember to turn off the pump (or disconnect the unit from the return line) for this procedure.

Suction-Side Systems

Suction-side automatic pool cleaners work off of the suction at the pool's skimmer. It is like vacuuming the pool without a pole or pool person. In this design, a standard vacuum hose of 1½-inch (40-millimeter) diameter is connected between the skimmer suction opening at one end and a vacuum head that patrols the pool bottom at the other end (Fig. 7-9).

The problem with these systems is that the skimmer was designed to skim dirt and debris from the surface of the pool before it sinks to the bottom. By using the skimmer's suction for another purpose, everything sinks to the bottom. Special valves can be installed on the skimmer to allow some suction to continue the skimming action while the remainder operates the vacuum head, but few pumps have enough suction to do both efficiently, so you end up with two systems that work poorly. Some valves alternate the suction between the skimmer and the vacuum head, but then you are back to the original problem that when the vacuum head is running, the skimmer is not.

The second major problem with these systems is that the debris collected is sucked directly into the pump strainer, often overloading the basket and cutting off circulation, making the entire system of filtration and heating inefficient. Canisters have been designed to float in-line along the suction hose, but when these clog, you are still slowing the circulation of your pool, which means it is not filtering or heating effectively.

Having mentioned the drawbacks, these systems are an excellent choice if your pool does not suffer from leaves or other heavy debris. Like other automatic cleaners, they are not meant to operate all of the time, so if you use them once or twice each week (or even a few hours each day), you will find them very effective and labor saving.

Like their boosterless water pressure counterparts, suction-side units are easy to install, repair, and maintain. Figure 7-11 shows a popular suction-side model made by Sta-Rite. Depending on the style of skimmer your pool uses, you will connect the automatic cleaner in one of three ways as shown. Note that in attaching the vacuum to the skim-

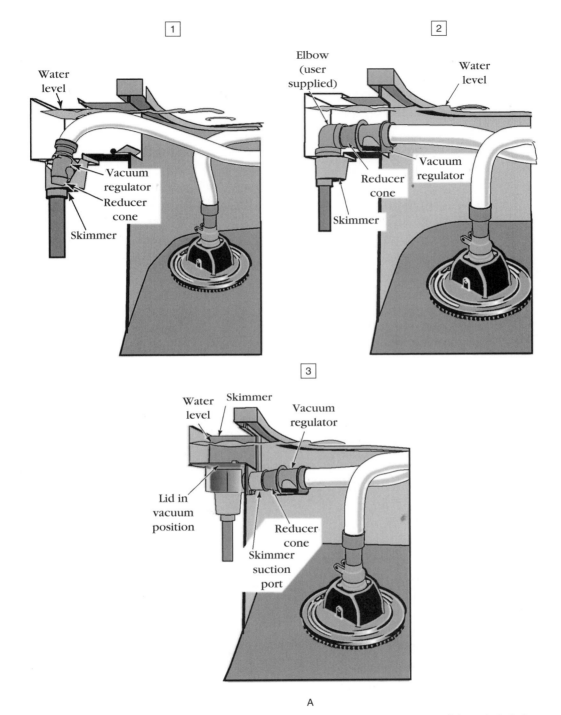

FIGURE 7-11 Suction-side automatic cleaner: (A) three connection methods; (B) exploded view and optional telepole. *Sta-Rite Industries.*

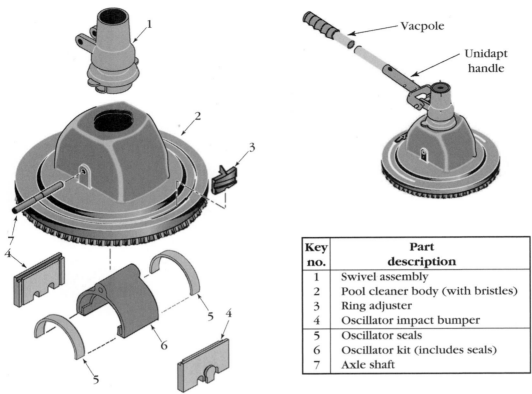

Key no.	Part description
1	Swivel assembly
2	Pool cleaner body (with bristles)
3	Ring adjuster
4	Oscillator impact bumper
5	Oscillator seals
6	Oscillator kit (includes seals)
7	Axle shaft

B

FIGURE 7-11 (*Continued*)

mer suction port (Fig. 7-11A, option 3), you drop your skimmer lid into the skimmer itself to force 100 percent of the suction into the port to maximize vacuum operation. If your pool is equipped with a main drain, close it off and direct all suction to the skimmer when using the suction-side automatic pool cleaner.

One of the best features of the suction-side model is the simplicity of parts (Fig. 7-11B), which are easily understood and replaced by any pool owner. The other unique feature is that the entire unit can serve as a manual cleaner when you want to vacuum the pool yourself.

Self-explanatory adjustments to the suction and turning radius of the unit can be made at the hose connection and on the unit itself. The only other common problem related to suction-side automatic

FIGURE 7-12 Hose kink problems.

cleaners is kinked hoses (Fig. 7-12). To avoid this problem, make sure the hose is not more than 6 feet (1.8 meters) longer than the diameter of your pool, and store the hose with as few coils as possible when it's out of the pool. Finally, try laying the hose out in the sun, stretched to its full length, for a few hours to relax any permanent kinks that may have developed.

To Buy or Not to Buy?

How do you decide whether to purchase an automatic pool cleaner in the first place? Well, here are some factors to consider.

PRICE

Obviously the cost of buying, installing, maintaining, and running the equipment is the first question (don't forget it costs money to replace parts as they wear out and costs more to run the cleaner itself, which is a good reason to use it no more than necessary and to keep the system pressure adjusted to specifications).

EXPECTATION

Make sure you understand what the various types of systems can (and cannot) do. Many pool owners buy a system and then expect the pool to be spotless every minute of every day thereafter. It just doesn't work that way, so understand that automated cleaners help, but they don't perform 100 percent of the cleaning duties.

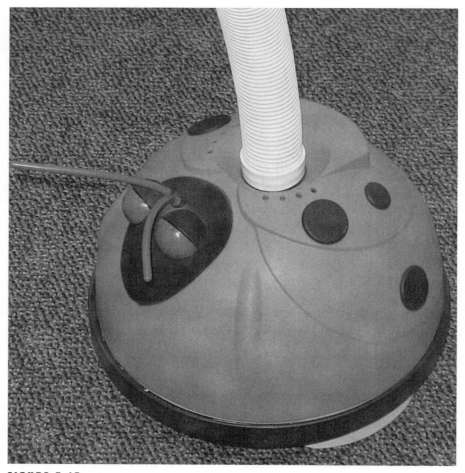

FIGURE 7-13 The "lady bug" model of suction-side auto cleaner.

EXISTING POOL AND EQUIPMENT

Will the pool actually benefit from an automatic system? Is it in such a heavily landscaped area that any system cannot keep up? Is the circulation system itself old and unable to do its part with an automatic system added?

SERVICE

If you have a service or clean the pool yourself on a regular basis (some people actually find vacuuming their own pool therapeutic), you may not benefit from an automatic cleaner. Then again, Olympic decathlon champion Bruce Jenner and his family named their vacuum head unit and treated it like a household pet, and many are designed to look like one (Fig. 7-13).

Lighting

External lighting of your pool is possible with a variety of high- and low-voltage systems. Remember that any lighting system that brings electricity near your pool must be inspected and maintained regularly to ensure safety.

The most common lighting systems for above-ground pools are over-the-edge models (Fig. 7-14). These are inexpensive, simple to install, and designed to plug into standard household current. Even though the actual current delivered to the light fixture is reduced from household voltage down to 12 volts, you should never attempt to service the unit before unplugging it from the power source. One unit will light pools of any size, and some are equipped with a rheostat to adjust the brightness of the light. Most models are designed for use in rigid-sided above-ground pools, not soft-sided.

This type of light fixture is designed to be cooled by the pool water, so never operate it unless it is fully submerged. When you submerge the unit (either for the first time, after replacing

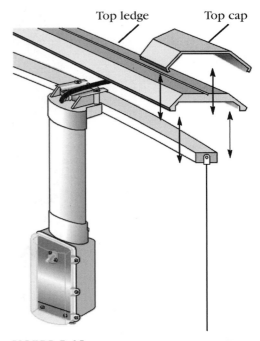

FIGURE 7-14 Above-ground pool light fixture. *SmartPool, Inc.*

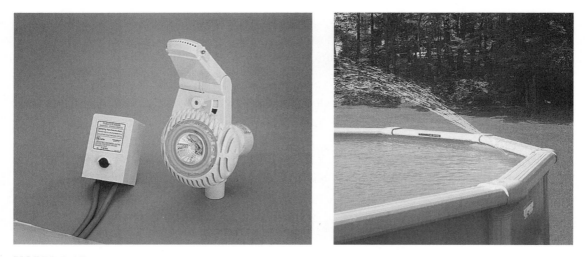

FIGURE 7-15 Light and fountain unit. *Zodiac American Pools, Inc.*

FIGURE 7-16 Porthole. *Delair Group, LLC.*

the bulb, or after reinstalling it in the spring), do so before connecting it to the electrical outlet. Look for bubbles that suggest the gasket might be leaking, flooding the interior of the fixture. If that happens, disassemble the unit, and clean or replace the gasket.

A clever variation of the over-the-edge light unit incorporates a fountain (Fig. 7-15) that is plumbed into the return line of the pool's circulation system. It also features an optional accent light that brightens the perimeter of the pool using fiberoptic technology. The fiberoptic perimeter light is designed to be attached to any above-ground pool, although some top ledges are designed to accommodate the lighting without additional adhesives. Fiberoptic lighting in this application is designed for an attractive accent only, not for illuminating the water itself. The fiberoptic cable derives its light from the pool light bulb that illuminates the water.

One more clever way to illuminate your pool is with a porthole (Fig. 7-16). These units require more work to install, although the process is similar to installing a skimmer to the pool (see Chap. 1). During the day, you can enjoy seeing swimmers from a unique perspective, and at night, any yard lighting will also illuminate the pool through the porthole. If you have weatherproof 12-volt yard lights near the pool, make sure they are set back far enough to avoid major spills or splashes. That said, outdoor low-voltage lighting is designed to withstand some sprinkles, so you can position one near the porthole for basic illumination of the pool.

Covers

Pool covers have come into increasing use over the past 10 years in an effort to save heating energy costs and water through evaporation. As water evaporates, the minerals in it are left behind in the remaining pool water. Chlorine, which is made from salt, leaves minerals in the water as well. If you live in say, Malibu, where the water comes from wells or other very hard sources, this is a very real problem. Covers slow down the rate of evaporation and, therefore, reduce chemical use. Covers keep out leaves and dirt which also absorb chemicals. Keeping out the dirt means you vacuum less and clean your filter less.

Shielding the water from the sun further cuts down on chemical use. The result is the water in the pool doesn't get so hard, so quickly. That saves you money in water bills, chemical costs, expensive draining, and, as if that weren't enough, the heater does not scale up inside from hard water.

Covers also help retain heat. Some estimates say a cover alone will heat the pool 10° to 15°F (5° to 7°C), meaning money is saved on gas or electricity. When you do turn on the heater, the water starts out warmer and heats up faster.

Cleaning the pool will be easier, and you will be happier because of the ecological, financial, and appearance savings on your pool. Okay, I'm sold. What are the choices?

Bubble Solar Covers

Have you ever unpacked something sent in the mail that is wrapped in that plastic bubble wrap? If you're like me, you squeeze the little bubbles between your thumb and finger to pop them, right? Well, someone realized that large sheets of this stuff would make good pool covers.

TOOLS OF THE TRADE: COVER INSTALLATIONS

- Razor knife or large scissors
- Tape measure
- Screwdriver set and loose-jaw pliers (if installing a roller)

Figure 7-17 shows a typical bubble cover (also called a *sealed air* or *solar* blanket). In profile, the cover has one flat side and one bumpy side. In fact, the cover is made from two sheets of blue plastic (usually available in 8 or 12 mil thickness), heat welded together with air bubbles in between.

The sun warms the air bubbles, which transfer the heat to the water (thus the term *solar cover*). Similarly, the trapped air acts as an insulator for the heat coming up from the water. Always lay a bubble cover on the water with the bubble side down. In this way, the spaces between the bubbles also act as pockets for trapped air, further insulating the water.

Because they are thin, lightweight, and flexible, bubble covers can be cut to any size and are sold in large sheets from 5 by 5 feet (1.5 by 1.5 meters) up to 30 by 50 feet (10 by 15 meters), with many interme-

FIGURE 7-17 Solar blanket cover and reel system. *FeherGuard Products Ltd.*

diate sizes to fit any job. They are easily cut with scissors or a razor knife. Bubble covers are cheap, costing about 25 cents per square foot (930 square centimeters). They last 2 to 4 years depending on water chemistry, weather conditions, and user wear and tear.

The disadvantage of bubble covers is that in heavy winds they can blow off or away into the next county. Also, they don't really keep out dirt or debris, because as you remove the cover the dirt falls into the pool. Much of the dirt does stay on the cover, meaning you have to spread it out somewhere and clean the cover as well as the pool.

On a large pool (especially if the cover has no roller), taking the cover off and putting it back on can be a real chore. If you drag it off, where do you put it? If you lay it in the grass or on a nearby deck, it might pick up more dirt that goes into the pool when you put the cover back on, or it might tear as you drag it back into place.

As the bubble cover ages, sunlight and chemicals make the plastic brittle, causing the bubbles to collapse and sending little bits of blue plastic into the pool and circulation system. In short, bubble covers are only good for their thermal properties, which are valuable, especially if you heat the pool a lot.

INSTALLATION

RATING: EASY

As you might guess, installation of bubble covers is very easy.

1. **Measure** Measure the pool and buy a cover that will overfit the water surface. In other words, get a size larger than you actually need.

2. **Lay Out** Lay the cover out on the water surface and leave it for 2 or 3 days (some manufacturers recommend as many as 10 days, but unless it is very cold, I have not found any difference after 2 or 3 days). You can remove it to use the pool during this period, but the idea is to give the material time to relax to its full size in the sun and to shrink to any degree that it might (I have found shrinkage is not more than 5 percent in any direction).

3. **Cut** Using shears, scissors, or a razor knife, cut the cover to the water surface size of the pool. As you walk around the cover, cut slightly less than you think you should—you can always go around again and trim off a bit more, but it sure is hard to add any back on if you cut too much. There are no frayed edges or seams to worry about.

ROLLERS

RATING: EASY

As noted, handling large bubble covers can be inconvenient. You can try to fold the cover into large folds before taking it off, which is easier with two people of course. You can also cut the cover into two smaller pieces, which might be easier to handle one at a time. Or you can buy a roller (also called a *reel system*). Figure 7-17 shows a typical roller. The concept is to attach the cover to the barrel of the roller with straps so that when you rotate the barrel, the straps roll up first, pulling the cover along and rolling it on the barrel.

Plastic covers are also sold to cover the cover. If left on the roller for extended periods, this cheap plastic cover keeps sunlight and dirt off the bubble cover, extending its life.

Installation of a cover roller is a simple matter of correct placement across the broadest diameter of the pool. Another style of cover management is a folder rather than a roller (Fig. 7-18), which is less costly and equally easy to use. The benefit of the folder is that you can easily remove the entire folded cover when using the pool.

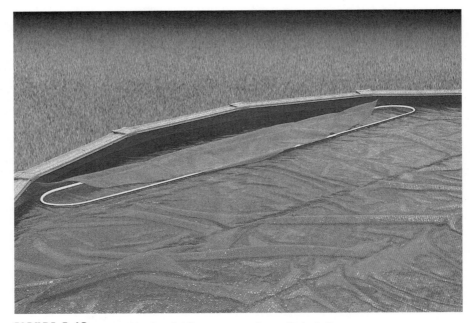

FIGURE 7-18 Solar blanket folding system. *Cantar/Polyair Corp.*

FIGURE 7-18 (*Continued*)

FIGURE 7-19 Dome cover.

Dome Covers

In some climates, you might want to use your pool while keeping it covered, protecting both the pool and swimmers from rain, snow, or other elements. Clear plastic dome covers are made for above-ground pools (Fig. 7-19). Dome covers are held aloft by a lightweight framework or, on larger pools, a blower that remains on at all times to keep forced air in the dome. These blowers are actually fairly efficient, requiring no more electricity than a 100-watt light bulb.

Most dome covers need some customization, so their installation is better left to a pro. Some are designed for specific pool makes and models, so ask your pool dealer if one is made for your particular pool if you think the dome cover is right for you.

Liquid Covers

A recent addition to the pool cover market is a liquid chemical that has the insulating properties of a fabric cover but doesn't require handling. The special liquid is actually alcohol and another active ingredient, which together form an invisible barrier on the surface of the water just one molecule thick. These liquids are said to be nontoxic, and they have been used in commercial pools for some years, mostly to reduce evaporation. Liquid

---

FAQS: OTHER EQUIPMENT AND OPTIONS

Why do I need a safety barrier or fence?

- Many localities require some kind of effective safety barrier to prevent children from gaining unsupervised access to the pool and possibly drowning. Whether required or not, a solid fence around your pool is a wise investment.

Isn't it dangerous to have electrical equipment near a pool?

- Yes, very dangerous. That's why all jurisdictions establish building codes that require setbacks for electrical equipment and outlets [most say a minimum of 5 feet (1.5 meters) away from the water's edge]. Low-voltage lights are another important safety investment, both for lighting your pool and any landscape lights adjacent to it.

Why can't I install a diving board on my deck?

- Diving boards are never permitted at any above-ground pool because even those with deep ends are not safe for diving or jumping. Slides are made for above-ground pools, but only to be used on pools designed for them and only to be used as intended (sliding, not diving or jumping).

Do I still have to vacuum the pool with an automatic cleaner?

- Yes. No automatic cleaner works 100 percent, since corners and other variations in shape cause the cleaner to miss some spots. Vacuums designed for above-ground pools have brushes built into them, so vacuuming also performs valuable brushing that keeps the liner clean (making it last longer) and free of algae.

Will a cover keep the pool warm?

- Yes. Covers may be all you need to keep your pool at swimming temperature all year around, especially if it is working to insulate heated water (from solar or gas heat). Evaporation also cools a body of water, and a cover will dramatically reduce evaporation.

---

covers are biodegradable and require renewal every 4 or 5 days, depending on weather factors and bather load. Of course, the liquid cover provides no protection from dirt or debris, so consider it only if your main concern is insulation and minor reduction of water evaporation.

Foam

Much like bubble covers, foam covers are sheets of lightweight compressed foam [1/8 inch (3 millimeters) thick] that float on the surface of the pool. Because foam is much more expensive (about $1.50 per square foot) than bubble plastic, these are primarily used for smaller pools. Installation is the same as with bubble covers, described previously.

Sheet Vinyl

Particularly in cold climates where it is desirable to cover a pool in winter, sheet vinyl is sold like a bubble cover without bubbles. Sheet vinyl has little insulating value and doesn't even float very well, but it is very cheap. Sandbags or plastic ballast bags that you fill with water are sold with sheet vinyl covers to anchor the cover on the deck or top rail as it lies over the water (Fig. 7-20).

Some building and safety codes require a pool to have a cover, on the theory that someone falling into a covered pool will not sink to the bottom as fast. These codes don't require you to use the cover, of course, so most of the vinyl covers I have sold were to satisfy the building inspector, then never taken out of the box. These covers are generally $10 to $20 for a 20- by 40-foot (6- by 13-meter) cover and can be purchased at any discount department store as cheaply as at your pool supply house.

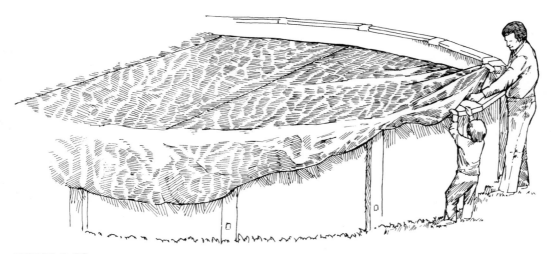

FIGURE 7-20   Sheet vinyl cover. *Cantar/Polyair Corp.*

FIGURE 7-21 Locking pool cover. *Loop-Loc Ltd.*

Mesh Covers

Where security is a real concern, you might use a plastic mesh fabric that stretches over the entire pool and anchors on the deck or top rail all around (Fig. 7-21). The middle part of the cover sags slightly into the water, and the material is made waterproof for this reason. The actual installation is a series of hooks and eyes that secure the unit in place.

These covers are strictly for security. They are a close woven mesh, like the mesh strips used on lawn chairs, reinforced with steel or other wires running in crisscross patterns within the fabric. Water can evaporate through these covers, although some water is trapped back into the pool.

A variation of the mesh security cover is the mesh leaf cover (Fig. 7-22 on p. 232), useful only as a means to keep large debris out of the pool, not for security or insulation value.

FIGURE 7-22 **Leaf net cover.** *Cantar/Polyair Corp.*

Water Chemistry

Let's begin with the easy part. See the sidebar on the next page for the components of healthy water (each of which will be explained in detail, so don't fret if it looks Greek to you now).

Oh sure, you'll get slight variations from one expert to the next even on these basics, but no one would completely disagree with these recommendations. In fact, if your pool tested out exactly within these parameters, you'd probably get a medal and your picture in a pool service magazine.

So if everyone generally agrees on the guidelines listed, where does the disagreement enter into it? The heated debate begins when you discuss how to achieve and then maintain these parameters; how to correct a severe imbalance of one or more of the components without throwing other components out of alignment; how to kill an unexpected algae bloom; how to eliminate blue staining from the plaster, etc., etc., etc. Well, here's my approach.

Demand and Balance

If you get nothing else from this chapter, learn this section on demand and balance—you will be ahead of many water technicians who think they know a lot about water chemistry.

BALANCED WATER

- **Chlorine residual: 1.0–3.0 ppm**
- **Total alkalinity: 80–150 ppm**
- **pH: 7.4–7.6**
- **Hardness: 200–400 ppm**
- **Total dissolved solids: Less than 2000 ppm**
- **Cyanuric acid: 30–80 ppm**

Water chemistry is a process of balance. Change one component, even to bring it into a correct range, and you might adversely affect another component, thus adversely affecting the entire pool. Imagine that the water quality parameters are stones, each of equal size and weight, evenly distributed around the edge of a dinner plate. Now imagine balancing that plate on one finger. You can do it if you find the exact center of the plate, where each stone balances the others. But now imagine that one stone is doubled in weight or removed. It changes the balance of the plate—the other stones will slide into new positions or off the plate completely, eventually making the entire plate fall from your finger and crash to the ground.

So it is with water chemistry. A balancing act, with each component working on the others. Moral of the story: learn about each component of chemistry, learn how to achieve ideal conditions for that component, but think before you pour chemicals. Learn about the consequences of each action (or lack of action).

This introduces the concept of *balance*, but what is the *demand* part? Water is a solvent. It will dissolve and absorb animal, vegetable, and mineral until it can no longer hold what it dissolves (the *saturation point*). After this, it will dump the excess of what it has dissolved (the *precipitate*). With this process in mind, understand that water makes demands on anything it comes in contact with until those demands are satisfied.

As you will see in the discussion of pH, for example, if the water is very acidic, it will demand to be balanced with something alkaline. If such water is in a plastered pool, the alkaline lime in the plaster will be dissolved into the water until that balance is achieved. When the water is no longer acidic, it will start dumping excess alkaline material, depositing it on tiles and inside pool equipment, as well as back on the plaster as rough, uneven calcium deposits. So demand and balance are inextricably related as will be increasingly apparent with each section of this chapter.

Components of Water Chemistry

Our original ideal parameters are worth repeating before I discuss each one individually.

- Chlorine residual: 1.0–3.0 ppm
- Total alkalinity: 80–150 ppm
- pH: 7.4–7.6
- Hardness: 200–400 ppm
- Total dissolved solids: Less than 2000 ppm
- Cyanuric acid: 30–80 ppm
- Temperature, wind, evaporation, and bather load

I've added one more component to the list. Obviously not a chemical component, temperature has an impact on some of the other components, either directly or indirectly, as do evaporation, wind, and dirt. More on that later, but for now just add it to your list.

Sanitizers: Chlorine

To discuss the first item on the list, chlorine residual, I will review the concept of sanitizers in general. So far in this book, when mentioning a sanitizing agent, I've only mentioned chlorine, but there are four forms of chlorine, each with its own properties and problems. There are many alternate sanitizing methods in common use, and there are a great number of "wonder" products for killing off special algae or solving stain problems. Chlorine is used as a benchmark because it is the most common sanitizer.

The purpose of any sanitizer is to kill bacteria in the water. Bacteria carry disease and stimulate algae growth. Sanitizers accomplish this by *oxidizing* the bacteria and other waste in the water. Rust is oxidization in progress. Oxygen is, in essence, dissolving the material with which it comes in contact. Sanitizers, oxidizing in your pool or spa are essentially "rusting away" the bacteria and other waste material in the water.

The most simple form of chlorine, which is found in nature as chloride mineral salts, is as a gas. It is made by passing electricity through a saline (salt) solution, one by-product of which is sodium hydroxide (caustic soda). Liquid chlorine (sodium hypochlorite) is manufactured by passing chlorine gas through this solution of caustic soda.

Dry chlorine is subsequently made by removing the water from such a solution. Before I examine the properties of each, let's look at some properties common to all, then some evaluation criteria for each form.

CHLORINE DEMAND

As I mentioned in introducing the ideas of water chemistry, water demands a lot. *Chlorine demand* can be defined as the amount of any chlorine product (in any form) needed to kill all the bacteria present in a body of water.

CHLORINE RESIDUAL

Chlorine residual is the amount of chlorine (expressed in parts per million) that is left over after demand is satisfied.

CHLORINE EVALUATION CRITERIA

Here are a few criteria for comparing various forms of chlorine and, for that matter, each of the other sanitizers I'll deal with. In each case, I will note any special properties of the sanitizer (for example, if the product is reasonably stable I will make no comment, but if it is very unstable I will say so).

Chlorine Availability: What does *available* mean—that you can ask it out for a date at any time and it will say yes? Well, sort of. Chlorine in its various forms combines with other elements present in the water, meaning some of the chlorine is locked up or unavailable for oxidizing bacteria, while some is available for that sanitizing purpose.

In considering the effectiveness of any sanitizing program in a body of water, this availability question must be considered to know what is really going on. In other words, you might add 5 gallons (19 liters) of liquid chlorine to your pool, but have less effective, active, sanitizing *available* chlorine than if you had added 2 pounds (0.9 kilogram) of chlorine gas. I will explain in a moment.

Stability: Chlorine is by nature very unstable, which is to say that it is easily destroyed by contact with ultraviolet (UV) light (sun). Efforts are made, as you will see, to make it more stable using various methods.

pH: Since you add some form of sanitizer to your water with such regularity, you must consider the pH of the product, because it will alter the pH of the water if enough is added.

Convenience: How easy is it to administer the product?

Cost: This is self-explanatory.

By-products: Does the form of chlorine you are using add anything to the water besides chlorine?

LIQUID CHLORINE

Liquid chlorine, sodium hypochlorite, has a high pH because of the caustic soda used in its manufacture. Although the caustic soda helps to keep the chlorine from escaping the solution, it gives liquid chlorine its high pH, between 13 and 14. Chemically, liquid chlorine is NaOCl.

The fact that chlorine is the product of salts makes liquid chlorine very salty. In fact, you are adding 1.5 pounds (0.7 kilogram) of salt to the water for every gallon (4 liters) of liquid chlorine you add. This adds to water hardness.

Liquid chlorine is produced at about 16 percent strength and by the time you use it, it is about 12.5 percent available. When you add sugar to your coffee, it tastes sweet. If you add twice as much, it tastes twice as sweet. If the object were to put as much sugar in solution in the coffee as possible, you would keep adding sugar until the coffee could simply not contain any more.

Why not do the same thing with liquid chlorine? Why not have, say, 65 percent chlorine? Then you would only need to carry 1 gallon

> ## TRICKS OF THE TRADE: CHEMICAL SAFETY
>
> One day in my pool supply store I was in one of those cleanup moods and thought I would combine two half-filled bottles of chlorine into one. What I didn't know was that one bottle contained muriatic acid and one contained chlorine. When I poured one into the other, holding them close in front of my face, the resulting explosion of chlorine gas (released by contact with the muriatic acid) choked me, burned my eyes and lungs, and I thought I was going to die. Fortunately I didn't, but the lesson was one I hope I can learn for you—don't make the same mistake.
>
> By the way, liquid pool and spa chemicals come in color-coded plastic bottles so, in theory, you can't make that kind of mistake. The color varies by manufacturer, but acid usually comes in red or green bottles while chlorine comes in yellow, light blue, or white bottles. Whatever the color scheme, don't ever do as someone in my shop did that nearly fatal day—don't ever put a different chemical in a bottle meant for something else.

of the product for every 4 gallons you now carry—it would be a more concentrated, convenient product. Unfortunately, because chlorine is so unstable, producing it at such strength is pointless. Within hours, that 65 percent solution would deteriorate to less than 15 percent anyway. So that's what you get from the factory to start with, ending up with about a 12 percent solution after some normal deterioration.

This brings me to another important point about liquid chlorine. Air, sunlight, and age accelerate the deterioration process, so keep your supplies fresh and covered. By the way, household laundry bleach, which is nothing more than liquid chlorine, is produced to around 3 percent strength. You realize how profitable the household bleach business must be when you note that a gallon costs about the same as a gallon of pool or spa liquid chlorine, yet the pool chlorine is four times stronger.

The advantages of liquid chlorine are that it is easy to use (just pour it into the body of water being treated) and it goes into solution immediately, because it's already a liquid. When pouring liquid chlorine, pour it close to the surface of the water to prevent splashing (bleaching your shoes) and to prevent unnecessary dissipation of strength (by contact with air).

DRY (GRANULAR OR TABLET) CHLORINE

There are essentially two types of dry chlorine sanitizers. The more popular of these two are called *cyanurates*—they contain stabilizer (cyanuric acid) to help prevent breakdown, generally about 1 pound (450 grams) of cyanurate to each 4 pounds (1.8 kilograms) of chlorine product. But more on that later. First, let's review the other type of dry chlorine.

Calcium Hypochlorite: Commonly sold under brand names including HTH, this product is popular for its relatively low cost and convenience of use. It is available in granular or tablet form.

Calcium hypochlorite is unstable, slow dissolving, and leaves substantial sediment after the chlorine portion of the product enters into solution in the water. It is 65 percent available chlorine, so it is rather potent (remember, liquid chlorine is only 12 percent available). Its pH is 11.5, so it tends to raise the pH of the water.

Calcium hypochlorite is often used as a shocking agent (see the discussion of shocking later). When shocking a pool, your main concern is available chlorine, delivered for immediate sanitizing use. Thus, a product that contains no stabilizer is preferred, because there is no point in paying the extra cost for chemicals that you don't need for this particular application. For regular sanitizing and for maintaining the chlorine residual, one of the cyanurates might be better because they are longer lasting with fewer by-products.

Cyanurates: There are two types of cyanurates, which are chlorine sanitizers containing stabilizer (cyanuric acid). The first form is *dichlor* (sodium dichloro-s-triazinetrione). Okay, I know I promised not to muck up this chapter with unintelligible chemical jargon and this product sure sounds like just that. But this is really simple. Most technicians refer to this product as dichlor. Triazinetrione just means stabilized—the product contains a stabilizing agent.

So what are the properties of dichlor?

- 56–63 percent available chlorine

- No sediment or by-products as with HTH-type products

- A pH of 6.8, so it is slightly acidic

When you buy dichlor products they won't be called dichlor. They will be called by various brand names, but read the label. Another good reason to read the label is to be sure the product is safe for vinyl-lined pools. Most dry sanitizers are harmless in small doses used for routine maintenance, but in larger quantities (for algae elimination or shocking the pool) they can discolor or even weaken vinyl fabrics.

The second form of dry chlorine is *trichlor* (trichloro-triazinetrione). Trichlor is the most concentrated (and therefore the most expensive) form of chlorine produced—90 percent available chlorine. To achieve both stability and strength, trichlor is produced mostly as tablets that slowly dissolve in a floater, although dry granular trichlor is produced as a super-killing agent such as Algae-Out for problem algae blooms.

Trichlor's pH is 2.8 to 3.2, very acidic. So make sure the floater (floating duck, dispenser, etc.) that you use in the pool is well away from any metal and/or skimmer or intake. Never put trichlor tablets in the skimmer as I have seen many pool technicians do. The acidic properties will have the same results as dumping acid directly into the system—the tablet dissolves, creating a few very acidic gallons of water. When the pump comes on, this acid water is sucked directly into any metal plumbing and components of the circulation system. This daily acid bath will quickly dissolve metals, destroying plumbing and equipment and depositing metal on the vinyl walls of the pool.

One major advantage of trichlor is that it dissolves slowly, so if a pool is not serviced frequently, chlorine will still be released into the water daily. A disadvantage of both dichlor and trichlor is that they are 50 to 58 percent cyanurate, so continued heavy use leads to a build-up of stabilizer, requiring at least a partial draining of the pool.

CHLORAMINES AND AMMONIA

As you might have guessed from the name, *chloramine* is a combination of chlorine and ammonia. Remember I said that chlorine likes to combine with other elements in the water. It is this fact that makes it an effective oxidizer of bacteria, as it tries to combine with the bacteria, thus killing them. But when ammonia is present in the water, the two will combine together to form chloramines.

You don't need any chemical formulas here. Just know these facts about chloramines:

- Chloramines are very weak cleaners (weak oxidizers of bacteria) because the chlorine is locked up with the ammonia and is not available to kill bacteria. The colloquial term is *chlorine lock.*

- Chloramines are formed when ammonia is present in the water from human sweat or urine (chemically very similar) and when there is insufficient chlorine present to combine with all the available ammonia and to oxidize bacteria. One active swimmer produces one quart of sweat per hour.

- Chlorine odors in a pool (and irritated eyes) are caused by chloramines, not too much chlorine.

- *Breakpoint chlorination* is the point at which you have added enough chlorine to neutralize all chloramines, after which the available chlorine goes back to oxidizing bacteria and algae instead of combining with ammonia in the water.

SUPERCHLORINATION

As the name suggests, adding a lot of chlorine to a pool is superchlorination. Why would you do it?

Even if you maintain a sufficient residual chlorine in your pool at all times, the ammonia and other foreign matter in the pool might keep your chlorine from being 100 percent available. That's why algae grows in a pool even though it has a high chlorine residual when you test the water. Superchlorination is, therefore, adding lots and lots of whatever chlorine product you use so that there is plenty in your water for the foreign matter to absorb, leaving enough to oxidize bacteria (kill algae).

Superchlorination is recommended by various experts as something to do monthly, three times a year, etc.—in other words, every pool person has a different opinion based on experience. The truth is that there is no one answer for every pool. You only want to superchlorinate when you have a substantial volume of ammonia or other foreign matter in your pool to soak up. Since you can't really know how much ammonia is in the water that might be locking up your chlorine, you can guess based on how dirty the pool gets every week and how much it is used. Of course, the more swimmers, the more sweat and urine are likely to be present. Therefore, it's a judgment call. Each chlorine product requires different amounts to achieve superchlorination. Remember, you have to close the pool to swimmers until the residual comes back into normal levels. So it pays to use enough, but no more. Follow the directions on the package for superchlorination. With liquid chlorine, I use 8 gallons (30 liters) of chlorine per 20,000 gallons (75,700 liters) of water. Keep the water circulating 24 hours a day until the residual reads normal.

Another time to superchlorinate is when the pool has been allowed to get exceptionally dirty or full of algae. A quick way to kill off a bad algae bloom (growth) is the same method I described to eliminate chloramines from your pool. Add ½ gallon (2 liters) of anhydrous ammonia and 4 gallons (15 liters) of liquid chlorine for every 10,000 gallons (38,000 liters) of pool water. In the chloramine discussion, I said that chloramines are not an effective cleaner. That was only half the story. Chloramines will, in fact, kill and dissipate

---

TRICKS OF THE TRADE: BREAKPOINT CHLORINATION

My method of achieving breakpoint chlorination is to add ½ gallon (2 liters) of anhydrous ammonia and 4 gallons (15 liters) of liquid chlorine to every 10,000 gallons (38,000 liters) of pool water and turn off the equipment for 24 hours. After that time, add another 4 gallons of liquid chlorine, turn on the equipment, and allow swimmers back into the pool when the chlorine residual reads normal.

Now this might seem odd—you're trying to use chlorine to eliminate ammonia, so why add more ammonia to the pool? Ammonia and chlorine combine to form chloramines. When ammonia decomposes, it forms nitrogen compounds, which are nutrients for algae. When the algae ingest the nitrogen, they also ingest chloramines, which kills them from the inside out.

quickly. They have a reputation as a poor cleaner because chloramines are totally unstable—they leave no residual, they kill and are gone.

So for especially bad algae problems, I use that method with the pump running 24 hours a day and continue to add chlorine daily [usually 4 gallons (15 liters) per day per 10,000 gallons (38,000 liters)] until the chloramines are absorbed and a normal residual is restored. The white powder you see in the pool the next day is the bleached, dead algae.

SUMMARY SCORECARD OF CHLORINE

To give you an overview of the chlorine products just discussed, here's a simple summary of their properties. They are arranged in order of relative cost, with gas being the cheapest. Note the relationship between cost and stability—the more stable you try to make this inherently unstable product, the more you must add to it and therefore the more expensive it becomes.

Product	pH	Available chlorine	Common form	Stability in water
Cl_2 gas	Low	100%	Gas	Very unstable
Sodium hypochlorite	13+	12.5%	Liquid	Unstable
Calcium hypochlorite	11.5	65%	Dry granular	Stable
Dichlor	6.8	60%	Dry granular	Very stable
Trichlor	3.0	90%	Tablet (or granular)	Very, very stable

In terms of relative effectiveness, this comparsion is useful: 1 lb chlorine gas = 1 gal sodium hypo = 1 lb, 8.5 oz calcium hypo = 2 lb, 13.5 oz lithium hypo = 1 lb, 12.5 oz dichlor = 1 lb, 2 oz trichlor, or 0.45 kg chlorine gas = 3.785 L sodium hypo = 680 g calcium hypo = 1.3 kg lithium hypo = 795 g dichlor = 510 g trichlor.

FORMS OF CHLORINE AND COMPARATIVE COST

Another important evaluation factor for which type of sanitizer to use is cost. The delivery system used, labor involved, transportation costs, and pH neutralizers needed (or other compensating chemicals required if the sanitizer chosen has "side effects" on the water as previously discussed) all must be factored in the real cost of the sanitizer.

Gas chlorine is the cheapest and most effective sanitizer, but it is also the most dangerous and difficult to use. To compare other sanitizer forms, assume we are comparing equal sanitizing action (not equal actual volumes or weights). Therefore, as a rule of thumb, here is a general guideline:

Liquid chlorine   costs 2 times the equivalent amount of gas

Calcium hypochlorite   3

Trichlor   3

Lithium hypochlorite   5

Dichlor   5

Sanitizers: Chlorine Alternatives

There are many alternatives to chlorine, although none are as widely used or as easy to use.

BROMINE

Bromine is a member of the chemical family known as *halogens*. In fact, so are chlorine, iodine, and fluorine. All are oxidizers. So if bromine is a cousin of chlorine, why isn't it used more? First of all, it cannot be stabilized and is therefore expensive to use. Having said that, it must be noted that bromine is more stable than unstabilized chlorine *at high temperatures*, and that's why many people use it instead of chlorine in their spas and hot tubs.

Nevertheless, a stabilized chlorine product is more effective and cheaper than bromine, so the only contribution made by bromine to the world of pools is for people who use their pool a lot and object to the chlorine odor. As noted in the section on chloramines, there won't be a chlorine odor if the correct amount is maintained and if the user doesn't urinate or sweat in the water.

The ultimate irony about bromine, however, is the fact that for it to work it must have a catalyst, which is usually chlorine. The small amount of chlorine added to bromine products is not detectable, so most technicians don't know that.

ULTRAVIOLET (UV) LIGHT

You know what ultraviolet light rays are: the rays that when they come from the sun turn your skin to fried chicken. While very effective at producing skin cancer, the problem with ultraviolet light as a sanitizer is that it is reactive, not preventive.

After the water has passed through your pump, filter, heater, and other equipment, it passes through a chamber where it is exposed to a beam of ultraviolet light. Any bacteria in the water passing through this light is killed. Do you begin to see the problem? The bacteria must be present in the water to be killed, whereas with a residual level of any chemical sanitizer already in the pool water, the bacteria is prevented from growing in the first place. Also, algae growing in the pool, which are not suspended in the water passing through the circulation system, are totally unaffected because the ultraviolet light has no way to kill the algae inside the pool itself. So ultraviolet light is not effective by itself. You still need to use chlorine or some other chemical agent to establish a residual in the water.

Another deterrent to UV systems is cost (typically over $700). One more thing. Don't confuse UV sanitizing systems with UV ozonators, which use ultraviolet light to create ozone, which is then injected into the plumbing system. Ozonators of several varieties are discussed next.

OZONE

Ozone consists of three atoms of oxygen and is created naturally in the earth's atmosphere by, among other things, lightning. It has been used for decades to purify municipal drinking and sewage waters and more recently for pool and spa sanitizing applications.

Ozone oxidizes contaminants in the water, but it is very unstable. Thus, it will react with contaminants immediately and break down, leaving little to attack algae on the walls of the pool. Therefore, a residual of chlorine, bromine, or another sanitizer is always required for a total sanitizing package. Of course, the purpose of using ozone is to reduce the amount of chemicals needed to achieve the minimum required residual.

For pools, ozone is created in one of two ways. One uses ultraviolet light (not to be confused with the ultraviolet systems described above) in a generator which contains UV bulbs that resemble tubular fluorescent bulbs. As air passes the bulb a photochemical reaction takes place that produces ozone. The benefit of the UV ozone generator is that it is cheaper than its counterpart (described next); the drawback is that it produces far less ozone.

The second method of ozone production is called *corona discharge* (CD). The CD ozone generator forms an electrical field (or "corona") which converts oxygen (O_2) to ozone (O_3), much like the creation of ozone by lightning. Generally, this produces 10 times more ozone for the same amount of electricity as a UV ozone generator, but the CD units are more costly because of the equipment needed to dry the air which passes through them. CD units require dry air because moisture combined with the electrical discharge can create corrosive nitric acid inside the appliance. CD units for small pools do not use an air drier component because of the volume of ozone needed to sanitize the smaller body of water, and they are thus much less costly.

Ozone has a neutral pH and generally has little impact on other parameters of water balancing. It is still necessary to shock the pool periodically as previously described. Other maintenance procedures are also unaffected, such as the need to brush and vacuum the pool, although ozone facilitates the flocculation of particles in the water, making them easier for the filter to remove.

Purchasing and installing an ozone system is easy (Fig. 8-1). There are several manufacturers who provide models based on the size of the pool and/or bather load. Each comes with plumbing and wiring directions which are easy to follow if you comprehend the plumbing chapter presented elsewhere in this book.

Ozone generators for residential pools have come a long way in the past few years and are a genuine and reliable product to assist with reductions in the use of harsh chemicals which many customers find objectionable. But remember, ozone is unstable and can't provide a sanitizer residual in the water, so it can only be used in tandem with chemical sanitizers.

BIGUANICIDES

A relatively new sanitizer most familiar by the trade name Baquacil, *biguanicide* is a general term referring to a disinfectant polymer that more accurately goes by the name polyhexamethylene biguanicide

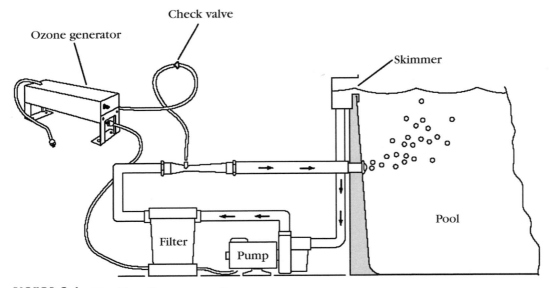

FIGURE 8-1 Plumbing of an ozone unit.

(PHMB). It is an effective sanitizer but not an oxidizer, so it cannot be used alone. A hydrogen peroxide product is applied as a monthly shock and a quaternary ammonium-based supplement is needed weekly. PHMB concentrations need to be kept between 30 and 50 ppm and can be tested with a special test kit using reagent drops. Other water balance parameters will be the same as for any other sanitizer.

PHMB cannot be mixed with chlorine products or, for that matter, any other chemicals except those designed as part of the package. The result if you do? Chocolate-brown-colored water and stains on the vinyl. This means if you change a chlorine-treated pool to PHMB, you must follow package instructions very carefully to first neutralize other chemicals and then remove any metals that may be present in the water. PHMB also reacts with household detergents and trisodium phosphate (TSP), so be careful with any other cleaning products in a pool that uses PHMB products.

The good thing about PHMB is that it has no equivalent to chloramines, so there is no objectionable odor or "combined" chemical problem to deal with. PHMB is a good product and a viable alternative which customers will appreciate, but more than with any other water treatment product, *read* and *follow* the label!

pH

pH is a way to assess the relative acidity or alkalinity of water (or anything else for that matter). It is a comparative logarithmic scale (meaning for every point up the scale, the value increases tenfold) of 1 to 14, where 1 is extremely acidic and 14 is extremely alkaline. In the middle, 7.0, is neutral, but for our purposes, 7.4 is considered neutral, neither acidic nor alkaline. In fact, the Los Angeles health department requires a pool's pH to be between 7.2 and 7.8.

pH has an amazing effect on water. If it is allowed to become very acidic (from adding too much acid, from too many swimmers sweating or urinating, from too much acidic dirt and leaves in the water) the water becomes corrosive, dissolving metal it comes in contact with. If

---

TRICKS OF THE TRADE: pH

- When you add acid, particularly liquid acid, pour close to the surface of the water so the acid doesn't splash on the deck or your shoes where it will do damage. Keep it away from your face and hands as well. If you get some on you, it might not burn right away, but don't be deceived—it will start eating your skin. Wash the area immediately with fresh water and seek emergency first aid if you are exposed to a lot or if you get it in your eyes.

- Also, when you add acid to the water, distribute it as much as possible, walking around the pool or adding it near a strong return outlet while the pump is running. The idea is to dilute it as soon as possible. Never add acid near a main drain, skimmer, or other suction inlet. The idea is not to have full-strength acid sucked directly into your pump and other equipment which could be damaged.

- When adding soda ash or other alkaline material, dissolve it in a bucket of water first and add it slowly. Adding too much or adding it too quickly might result in the water turning milky blue. Follow the package directions. By the way, soda ash has a pH of 12 and bicarb has a pH of 8. These products are described later, but for now, use soda ash to raise pH and total alkalinity. Use bicarb to raise total alkalinity alone.

- In both cases, when adding acid or alkaline, allow several hours for the pH to stabilize before testing again—the next day is even better. If you test right after adding chemicals, you will get a false reading. Similarly, don't test pH after adding chlorine or other sanitizers. Remember, each has a significant pH itself, so if you test right after adding sanitizer, you might be testing water saturated with that product and not truly reflective of the water's pH.

it is allowed to become too alkaline, calcium deposits (scale) can form in plumbing, equipment, and on pool walls.

More than that, pH determines the effectiveness of the sanitizer. For example, you might have a chlorine residual of 3 ppm in a pool with a pH of 7.4. But increase the pH just four-tenths to 7.8 and you will still have a chlorine residual of 3 ppm, but the ability of the chlorine or bromine to oxidize bacteria and algae will decrease by as much as one-third. Therefore, maintaining a proper pH is not just a factor of interesting chemistry or saving your equipment, it is a matter of effective maintenance.

pH is adjusted with various forms of acid (to bring down a pH that is too high) or alkaline substances (to raise a pH that is too low). With plaster pools, the calcium in the plaster creates a situation where the water is always in contact with alkaline material, slowly dissolving it (remember, water is the universal solvent), causing the water to be slightly alkaline. Therefore, in plaster pools you will always be adding small amounts of acid to keep the pH down. Above-ground pools create the exact opposite environment because water is always in contact with nonreactive surfaces such as plastic liners. Water can easily become too acidic from bather load, dirt, and other factors, so acid neutralizers must be added frequently to keep the water from destroying metal components in the heater or other equipment. Raising the pH is normally done by adding soda ash or specialty liquid products. Follow package instructions and be careful not to add too much, which can create just as many problems as you are trying to solve. At times, even above-ground pools can become too alkaline, from substances in the sanitizers or stabilizers you add.

Acid is most readily available in liquid (a 31-percent solution of muriatic acid) or dry granular forms. As noted elsewhere in the book, never mix acid and chlorine—the result is potentially **deadly** chlorine gas.

Because each product is different, I won't try to guide your use of acid; just follow the instructions on the package or in your test kit for the amount to add. Keep this general guideline in mind, however. If you add acid in small amounts, you can always add a bit more to get the desired results. But if you add too much, the water becomes corrosive, which means you must now add alkaline material to bring the pH back up.

pH is tested using a chemical called *phenol red*. For easy reference, all discussion of testing techniques is in the section on chemical testing.

Total Alkalinity

Total alkalinity is one of those concepts that frequently gets confused with pH or hardness. While the total alkalinity of a pool has an impact on those and other factors, it is not the same thing.

Imagine a waiter brings you a cup of coffee already sweetened with sugar. You sip the coffee and rate the flavor as slightly sweet, moderately sweet, very sweet, very, very sweet, and so on. But does that rating tell you how much sugar was actually put into the cup? To know that, you would ask the waiter and he might tell you, for example, it was 1 tablespoon (15 milliliters).

Total alkalinity is to water chemistry what that tablespoon of sugar is to the cup of coffee; while pH is to water chemistry what the flavor rating is to that coffee. A pH test will tell you the relative acidity or alkalinity of the water, while the total alkalinity will tell you the quantity of alkaline material in the water. It is a measurement of the soluble minerals present in the water (like the measurement of sugar in the coffee).

In other words, *total alkalinity* is a measurement of the alkaline nature of the water itself and, therefore, the ability of that water to resist abrupt changes in pH. In fact, adjusting water chemistry to a proper total alkalinity acts as a buffer against fast and extreme changes in pH (called *spiking*).

The actual test method is discussed later, but it is important to know at this point that an appropriate reading for your water's total alkalinity is 80 to 150 ppm. Less than 80 ppm means that too much acid has been added even if the pH reading is high. Therefore, you always adjust the total alkalinity level first, then the pH.

Proper maintenance of the total alkalinity of a pool will pay great dividends. You will use less chlorine, less acid, and see fewer algae problems.

Hardness

Hardness (or calcium hardness) is a component of total alkalinity. It is a measurement of the amount of one alkaline, soluble mineral (calcium) out of the many that might be present.

So why single out this one alkaline mineral for special measurement, especially if you have measured and balanced total alkalinity? Because, in sufficient quantity, calcium readily precipitates out of solution and forms salty deposits on the pool and in equipment. This deposit is called *scale* and is the white discoloration you see at the waterline of a pool. If you live in an area with hard water (water that contains large amounts of minerals) you might also have seen it in your home—in drinking glasses or flower pots that are left with standing water for long periods.

The acceptable range when testing for hardness is 100 to 600 ppm. Over 600 ppm and you will see scale. The only cure for water that hard is to partially or totally drain the pool and add fresh water.

Total Dissolved Solids

Now that I've discussed two measurements of minerals in the water, there is one overall category that helps you keep track of the big picture. *Total dissolved solids* (TDS) is, as the name implies, a measurement of everything that has gone into the water and remained (not been filtered out), intentionally or not.

The total of minerals (including calcium), cyanurates, chlorides, suntan lotion, dirt, etc., etc., equals TDS. The main contributor to TDS increases in a pool is evaporation. When the water evaporates, it leaves its contents behind. You add more water, and the solids keep building up as it evaporates. As noted earlier, the liquid chlorine you add to the pool is made from salt, so when the liquid evaporates the mineral remains behind, adding to the total amount of dissolved solid material in the water.

In most places, water from the tap already has a TDS reading of 400 ppm. You add about 500 ppm more each year with chemicals. Depending on where you live, the evaporation rate might add another 500 ppm per year. In California, for example, between wind and temperature you can expect to lose the equivalent of the entire volume of your pool each year to evaporation.

When a body of water reaches a reading of 2000 to 2500 ppm, it's time to empty it. There is no other way to effectively remove all those solids floating around in your water. Of course, if you live in a cold climate, where you drain all or part of your above-ground pool each winter anyway, this will be less of an issue.

If you can't see, taste, or smell TDS, why make such a big deal about it? Because TDS acts like a sponge, absorbing chlorine and other chemicals you put in the water, rendering them ineffective or requiring you to use far more of each chemical to attain the same results. That becomes expensive and creates further problems (remember, balance).

Conditioner (Cyanuric Acid)

Cyanuric acid simply extends the life of chlorine in water by shielding it from the ultraviolet rays of the sun, which would otherwise make the chlorine decompose. Cyanuric acid is a powder that is added once or twice a year in most pools, maintaining a level of 30 to 80 ppm. Above 100 ppm serves no function and creates other chemistry problems.

Because cyanuric acid helps prevent chlorine decomposition, it is also frequently called *stabilizer*. Because it extends the life of the chlorine, the result is that you use less chlorine and save money. You will thereby be adding less mineral to the water, slowing the rate of increase of TDS.

Cyanuric acid itself does not decompose or burn out in pool or spa water as is commonly believed. The only way to dissipate it is to remove it through water replacement. Much is also lost when bathers splash and leave the pool and when removing leaves, dirt, or other debris during cleaning.

As with the other components of water chemistry, the testing methods are detailed later. The method of administering cyanuric acid to water is detailed in the chapter on cleaning and servicing.

Weather

Perhaps the least discussed and yet most important aspect of water chemistry is the impact of varying weather conditions. Although you can't do much about these factors, understanding the role they play in your pool will help you make decisions about adding water and chemicals and whether or not to use a cover.

SUN

As mentioned, ultraviolet rays in sunlight speed the decomposition of chlorine. Without stabilizer, 95 percent of chlorine can be lost in two hours on a sunny day. In addition to protecting chlorine with stabilizer, a pool cover might pay for itself in the chemical savings over many years. Similarly, wind, low relative humidity, and high temperatures will speed evaporation and, as I have just discussed, that increases the frequency of draining your pool.

TEMPERATURE

In Malibu, pool water will stay in the 70s (near 21°C) during the summer and below 60°F (15°C) in winter. Although unscientific, I have found that below 65°F (18°C) you will have little or no problem with algae growth. Above that temperature you are always fighting it.

WIND AND EVAPORATION

Wind not only speeds evaporation, but it also carries dirt into the pool, adding to the TDS (not to mention adding to your cleaning problems). Evaporation becomes a real problem when you see your water bill going up and begin to question if there is a leak in the pool.

DIRT

The dirt, leaves, and other debris carried into a pool by wind also impact on chlorine use. Such debris will absorb chlorine, effectively removing it from the water. You will need to add two to three times as much chlorine to a dirty pool than you will to a clean one to reach the same results. One more reason to keep the pool clean, as well as the skimmer basket, strainer basket, and filter.

RAIN

Rain is another weather problem. In some areas, such as Southern California, the rain comes in enough volume during certain months that you have to pump water out of the pool to prevent overflowing and flooding. In doing so, your carefully balanced chemistry is quickly unbalanced. The extra water saves your water bill, but plays havoc with maintaining chlorine residual and other levels, requiring far more tweaking of water chemistry during rainy months than dry.

In summary, it is important to realize that varying weather conditions and bather loads have varying impacts on your water chemistry. Simply being aware of that probability will keep you testing and adjusting your water frequently enough to stay ahead of rapid changes that can be caused by weather influences.

Algae

The best cure for algae is keeping a clean pool, filter, and baskets. Keep your chemical components balanced and brush the pool frequently, even if there is no apparent algae or dirt. The brushing exposes microscopic algae growth to sanitizer before it has a chance to really blossom, preventing a major problem.

If you do encounter a situation of advanced algae growth, there are two approaches to deal with it. One is a general elimination program that will work with most algae blooms. The other is to combine that program with a special algicide. Please note, however, that there are no miracle cures. Despite the claims of some manufacturers, there are no chemicals you can simply add to the water and walk away, expecting the algae to disappear by the next service call. Every elimination technique requires hard, repetitive work—another good reason to take every preventive measure in the first place.

General Algae Elimination Guidelines

RATING: EASY

The following steps will help with any type or spread of algae growth and should be undertaken before adding algicides.

1. **Clean the Pool** Dirt and leaves will absorb sanitizers, defeating the actions you are about to take. Be sure you have a clean skimmer and strainer basket and break down the filter as well, giving it a thorough cleaning. As the algae dies or is brushed from the pool surfaces, it will clog the filter, so it is important to start clean. You might need to repeat this process several times during a major algae treatment.

2. **Check the pH** Adjust it if necessary (see water treatment section).

3. **Sanitize** Add 1 gallon (4 liters) of liquid chlorine for every 3000 gallons (11,000 liters) of water. Brush the entire pool thoroughly, stirring up the chlorine for even dissolution and distribution. Never use trichlor or other granular sanitizers at such strength on vinyl-lined pools. The surface will discolor.

4. **Circulate** Run the circulation for 72 hours, allowing the chlorine to attack the algae at full strength, brushing the pool at least once each day. Adjust the pH as needed and continue to add liquid chlorine to maintain a chlorine residual of at least 6 ppm.

5. **Filter** As noted earlier, the dead algae might clog the filter, requiring teardown and cleaning several times. Keep an eye on the pressure. When the chlorine residual returns to 3 ppm, resume normal

TRICKS OF THE TRADE: SPOT ALGAE REMOVAL

- Use a length of 2-inch (50-millimeter) PVC pipe. Insert the pipe into the water, with one end over the algae. Pour a cup of chlorine into the other end of the pipe and allow it to settle over the spot.

- Don't use dry chlorine or tablets for this procedure on vinyl-lined pools. The concentrated dose will discolor vinyl and can weaken fabrics. Some new forms of dry sanitizer are safe for vinyl-lined pools, so always read the label.

maintenance. You will need to vacuum the pool frequently during this period to remove the dead algae, which will appear as white dust when you brush or otherwise disturb the bottom.

You will also need to brush frequently, not only at first, but for at least one week after you can no longer see any trace of algae. Believe me, it's still there and will rebloom if you let up.

If these measures are not effective on your particular situation, or for more certain, rapid results, you can also use an algicide.

Algicides

Algicides kill algae, while algistats inhibit their growth. In fact, most products can be used in varying strengths for each purpose. There are countless companies producing countless products designed to prevent and/or kill algae in pools. Brand names vary and because many of them use a chemical as part of their name, they might be confusing.

COPPER SULFATE (LIQUID COPPER)

Effective on all types of algae, but recommended for use in dark-colored pools because of staining.

Copper sheets were nailed to the bottom of sailing vessels for centuries to eliminate marine organism growth, such as barnacles. It remains a leading ingredient in paints applied to boat hulls. In the last hundred years, copper sulfate (also called *bluestone*) has been used as an algicide in lakes and decorative ponds.

The copper actually destroys the algae's ability to breathe and consume food, killing it in the process. It also tends to coat the pool surfaces, acting as a preventive long after the original problem it was applied for has disappeared.

The only drawback to this product is that it stains some pools, along with swimmers' hair, that distinctive blue-green copper color you see on buildings that have copper roofs or trim. Modern products containing copper sulfate also contain other chemicals (called *chelating agents*) to prevent the staining, but in the simultaneous presence of strong oxidizers, like chlorine, the staining returns. Products containing copper sulfate are extremely effective, but I only use them on dark-colored pools where the potential stains will go unnoticed.

COLLOIDAL (SUSPENDED) SILVER

Effective against all types of algae in all situations.

Silver works the same way as copper, without the side effects. I have heard of extreme cases where silver, which is sensitive to light, has turned black in a pool if administered in sufficient quantity and catalyzed by the presence of excessive amounts of acid or stabilizer. In practice, I have never seen this occur and find products such as Silver Algaedyne to be remarkably effective.

The first time I used Algaedyne I followed the label instructions and poured a quart into the skimmer of actor Dick van Dyke's pool. In a minute, the return lines of the pool were spouting forth a brown liquid akin to motor oil. I was horrified because the label failed to warn me about this temporary effect. Fortunately, it dissipated in a few minutes and actually allowed me to dramatically observe the circulation pattern in his pool. The moral of the story is to follow label instructions, but be aware that they might not tell you the whole story. When you buy a new product, ask the salesperson what to expect.

POLYMERS

Effective on all types of algae to some degree, but works best on green and yellow algae.

Polymeric algicides might be any of a number of chemical compounds, preceded by the prefix *poly*, meaning many parts which duplicate each other. In short, they work by repeating the job over and over.

The way polymers accomplish the job is to invade the outer membrane of the algae and effectively smother it. Polymeric compounds contain a strong, positive electrical charge and are attracted to algae, which are naturally negatively charged. Herein lies the drawback. They are also attracted to dirt in the water or filter, so their strength might be diluted.

Also, because they work by attacking the exterior of the plant, they're not very effective against black algae, which have a very strong outer shell, or yellow algae, which have a stronger exterior than green, unless you have done a complete job of brushing (which might be impossible in some pools). Also, polymers don't coat pool surfaces and build much preventive effect. In other words, use them on green algae problems or where you're sure you have brushed thoroughly, but not with any hope of preventive maintenance. When you do use them, apply 20 percent more than the label suggests to allow for dirt in the pool.

QUATERNARY AMMONIUM COMPOUNDS (QUATS)

Good on green algae in early stages. Quats work like polymers, but the positive electrical charge is not as strong. Generally I have been unimpressed with the results of these products. Further, many of them will foam up if agitated, but filtration will ultimately remove the foam.

CHLORINE BOOSTERS

Great against stubborn yellow or large green algae blooms. The most effective chlorine booster is anhydrous (or aqua) ammonia. As described previously, it turbocharges the chlorine so it is more effective and kills algae from literally the inside out. It is good and inexpensive, especially for large pools.

Common Water Problems and Cures

We have reviewed the most common problem associated with pool water—algae—and now let's take a look at the symptoms you might see with other common problems.

Cloudy Water

Cloudy water is most often caused by inadequate filtration, either signifying that the daily filter run is not long enough or the filter itself is dirty. Solutions to those two problems are obvious. DE leaking through a filter back into the pool can also cloud the water, so check filter grids for holes or seams that have separated.

Finally, check the total alkalinity and pH. Sometimes high readings result in cloudy water. In spas, this might reflect high TDS levels in a relatively confined, small body of water. Adjust as described later.

Colored Water

If the problem is not algae, color in pool water means metals. It also means problems with staining, so make corrections promptly.

GREEN

Green or blue-green means copper. The water is probably etching (very acidic), so balance the pH. The copper might have come from plumbing or heater components, stripped away by acidic water and deposited on light-colored vinyl walls. It can become very extreme and actually appear as black, sooty-looking deposits. When this happens, immediately shut down the circulation equipment until you can ascertain the problem and solution. Apply the corrective measures and distribute the chemicals by brushing the pool. After several hours, take another pH reading. If it is now normal, turn the equipment back on; otherwise, keep balancing. The copper might also come from a copper-based algicide, so examine the other products you might have used to eliminate any sources.

Finally, apply a metal chelating agent to the water. These products, available in various brand names, will attract and combine metals so they can be filtered out. Follow the label instructions on the product you choose and in all cases, run the filter for 72 hours once you have balanced the chemistry.

BROWN TOWARD RED

This is caused by iron. Check and adjust the total alkalinity first, then the hardness, and finally the pH. You might need to drain part or all of the water and add fresh. Iron precipitates out of the minerals in the water, built up over time, from corroding light fixtures, ladders, and metal rails or from plumbing corroded by acidic water. Add a chelating agent to the water.

BROWN TOWARD BLACK

This is caused by manganese or other heavy metals (including silver). Follow the same instructions as for iron. Remember, the precipitated metal might have come from another chemical product you were using for algae control or some other purpose, so examine what has recently gone into the pool. Usually, however, the cause of all metal staining is water that is too acidic.

Scale

As described previously, scale is a buildup of calcium carbonate precipitated out of water by evaporation or heat. Of course, excessive amounts of calcium need to be in the water in the first place for this formation to occur.

The solutions are simple. Check the hardness. If it is near or exceeds the standard of 200 to 400 ppm, drain and replace some or all of the water. If the hardness is within acceptable limits, the problem might be high pH or total alkalinity. Check both.

Scale can be a frequent problem in small pools where high bather loads cause heavy use of chemicals and high rates of evaporation. The water quickly becomes hard. Be prepared to change water in a small, heavily used pool on a regular basis.

Eye or Skin Irritation

Irritation of the eyes or skin can be caused by too much chlorine, although that is rare except perhaps if you are too heavy-handed with your chlorine. Most such problems can be traced to low pH and/or too many chloramines. When the free, available chlorine is insufficient to oxidize the chloramines, you will have the characteristic chlorine odor and eye or skin irritation complaints. The solution is to adjust the pH and shock the pool.

Colored Hair, Nails, or Skin

Caused by the same problems as eye or skin irritation, discolored hair, nails, or skin can be a real problem. The most susceptible are light-skinned, blonde persons.

Odors

There are several odors you might use as clues to pool problems.

CHLORINE ODOR

See the section earlier on eye and skin irritation.

MUSTY ODOR

Algae growth or high bacteria in a small pool can lead to a musty, moldy smell. The solution is to eliminate the algae as described earlier, shock-treat the water to kill any bacteria, or simply drain and refill it.

A smell that is closer to mildew might, in fact, be mildew on pool covers or in deck crevices where water has been standing for long periods. The cure is to follow your nose to the source and take corrective measures.

Stains Unique to Vinyl-Lined Pools

Vinyl-lined pools will suffer from many of the same staining problems as any pool, but they are also subject to one unique type. Mold can appear as a brownish-black stain that is actually the spores growing up from the moist ground through the liner itself. These molds are harmless to swimmers or the liner, but are simply unsightly.

High chlorine residuals (around 5 ppm) will bleach them out in most cases, but that may weaken the liner itself. The only real cure for this mold problem is to dismantle the pool (which you may do in autumn anyway depending on the climate), and before setting it up again, lay down a thick polyethylene sheet (available at any hardware store) as a pad for the entire pool. Mold will penetrate vinyl, but not poly.

Water Testing

Having reviewed various potential problems of pool water and learned that diagnosis is based on testing the water, I will now examine the methods of doing so. Each component of water chemistry has unique test requirements, but the most popular basic methodology is similar.

Test Methods

There are three basic approaches to water testing: using chemical reactions and comparing the resulting colors, estimating values with electronic devices, and making observations of the relative cloudiness (turbidity) of a water sample.

COLORIMETRIC

The most common approach to testing water is to collect a small sample in a clean tube and add some chemicals to the sample (Fig. 8-2). These chemicals, called *reagents*, let you evaluate the sample by its changing color. By comparing the intensity of the color with a color chart of known values, you can determine the relative degree of each water chemistry parameter.

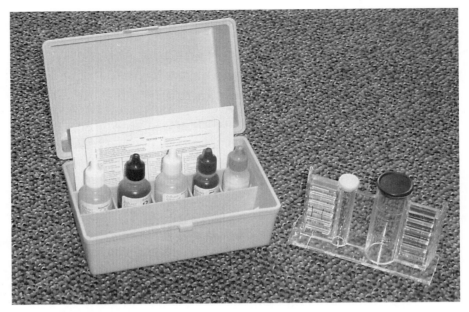

FIGURE 8-2 The typical pool combination test kit.

TITRATION

Also a color-based test, *titration* is the process of adding an indicator reagent (a dye) followed by a second reagent (called the *titrant*) in measured amounts, usually a drop at a time, until a color change occurs. By counting the number of drops of reagent required to effect the color change, you can estimate the value in question.

TEST STRIPS

Paper strips impregnated with certain test chemicals are also used as colorimetric test media. With this method, you dip a test strip into the pool and move it around in the water for 30 seconds. When you remove the strip, you compare its color to the color chart of known values to determine the test results.

ELECTROMETRIC

Testing using electronic probes attached to calibrated digital or analog readouts is the most accurate method of chemical testing (Fig. 8-3).

TURBIDITY

In a turbidity test, a vial is provided with a dot on the bottom. A water sample is taken and combined with a reagent called *melamine*, which clouds the sample. When viewing the sample from above, you compare the appearance of the dot with a chart of known values to determine the results.

Let's examine each parameter of water quality and identify which test is used for each.

Chlorine Testing

Several reagents are used for testing not only the presence of chlorine, but also the free, available chlorine within the total.

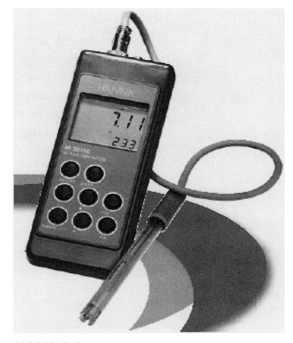

FIGURE 8-3 **Test meter.** *Hanna Instruments.*

OTO

The first bottle in most reagent test kits is OTO, or more exactly, orthotolidine. When combined with a sample that contains chlorine, OTO (which contains muriatic acid) turns color, from light pink to yellow to deep red, depending on the strength of the chlorine. You compare the color of the sample with the color chart to determine the exact parts per million of chlorine in the water.

OTO can also measure bromine in the same way, using a slightly different color chart and with similar limitations. (Multiply the chlorine reading by 2.25 to get the bromine ppm.) Bromine results will turn a darker yellow.

DPD

A second colorimetric test, which determines the actual free chlorine in a sample, is conducted with a reagent called DPD (diethyl phenylene diamene). By subtracting the results of the DPD test from the OTO test, you will know the amount of combined chlorine—chloramines—in the sample.

pH Testing

pH is also tested with reagents and a color comparison chart. The reagent for this test is phenol (phenosulfonephthalein) red. Before adding phenol red to a sample, you must neutralize any chlorine in the sample to avoid a false reading, and add a predetermined amount of the neutralizer sodium thiosulfate.

Perhaps the best use of electronic testing devices is in determining pH. They are accurate, and they are not subject to problems from high chlorine or bromine levels.

Total Alkalinity Testing

The most common chemical test for total alkalinity of a water sample is the titration method described earlier. There are no significant limitations in total alkalinity titration tests such as those described for other reagent tests.

Test strips are available for total alkalinity testing. With both pH and total alkalinity testing, there is one test to determine the relative value (what is the pH and what is the total alkalinity value?), followed by another test to determine the acid or base demand. In short, a second titration test determines the amount of acid or alkaline (base) material required to bring the water back to desired levels.

Hardness Testing

Hardness is tested using a colorimetric and titration test combined or test strips as described previously.

Total Dissolved Solids and Heavy Metals Testing

There is no simple, inexpensive way to test for total dissolved solids or heavy metals in a water sample, so take a sample to a lab or local pool retailer that is equipped with electronic equipment to analyze for these parameters.

Cyanuric Acid Testing

Cyanuric acid is evaluated with the turbidity test described earlier. There is an electronic device available, but it is expensive. Also, you can use test strips that have recently been added to the water testing marketplace.

General Test Procedures

RATING: EASY

Regardless of the type of testing you are conducting, some basic precautions and procedures are necessary.

Use only fresh reagents or test strips. Keep them out of prolonged direct sunlight and store them at moderate temperatures. Look at reagents before use. If they have changed color, appear cloudy, or have precipitate on the bottom of the bottle, discard them.

Thoroughly rinse the sample vials and any other testing equipment using the water you are about to test. Never clean equipment with detergents because chemical residue from these products can deliver false readings.

Consider where and when you are sampling to ensure truly representative results. Samples should be collected several hours after any chemicals have been added and thoroughly circulated through the body of water to ensure that their effect has been completely registered. The water should be circulated for at least 15 minutes before sampling. Sample away from return outlets and away from dead zones in the pool to make sure you are collecting a representative sample. There are times, however, when you might want to know more about various zones in the pool, to evaluate circulation for example, in which case you might specifically test these areas. Finally, collect samples from at least 18 inches (45 centimeters) below the surface to avoid inaccuracies caused by evaporation, direct sunlight contact, dirt, etc.

Make color comparisons in bright white light against simple white backgrounds to ensure accurate colorimetric comparisons. Most test kits come with a white card to place against the back of the test vial.

Observe and record the results immediately; never take samples for later evaluation. The chemistry of the water can change when a sample sits around.

Handle samples and reagents carefully. As noted earlier, some reagents can impact health, so avoid direct contact and never pour them into the pool or spa, even after the test is complete. Avoid contact with the sample because acids and oils from your hands can contaminate a sample, leading to false readings.

TRICKS OF THE TRADE: WATER TESTING

- Take your time. Hurried testing leads to inaccurate results, which leads to improper application of chemicals. Follow the directions in the kit. Some might seem redundant or unnecessary, but believe me, they are stated for a reason, so follow them. Also, allow time for hot samples to cool for more accurate results.

- Never flash test. Old-time pool and spa technicians often carry a bottle of OTO and one of phenol red in a leather case strapped to their belt. They arrive at a pool and dash a couple of drops of each into the pool, saying they can analyze enough of the chemistry from that. If you have followed the preceding discussion of the many factors influencing the accuracy of various test methods, you will know that this is simply false. On top of the many reasons for more careful analysis techniques, there is a very different color appearance when viewing a dispersing reagent in the pool as compared with a captured result in a vial held at eye level.

- Conduct all tests before adding any chemicals to modify the water chemistry. If you test for chlorine residual and add significant amounts of chlorine in any form, then test the pH, your pH reading will reflect the value of the chemicals just added, not the water. After adding chemicals to the water, allow adequate time for distribution, then test to see if your actions were adequate. Allow at least 15 minutes for liquid chlorine to circulate and at least 12 hours for the pH to adjust before testing again.

- Because test kit reagents can be replaced, you will probably use your kit for many years. The color chart might, however, fade over time. Compare your chart against a new one from time to time, or simply buy a new test kit annually (they're not that expensive).

Water Treatment

I have reviewed the various testing methods to determine the chemical needs of a body of water, but what about administering the actual chemicals that will deliver the desired results? There are as many products for sanitizing and balancing the chemistry of water as there are for testing it. It pays to follow the directions on product labels, but some general guidelines apply.

Liquids

RATING: EASY

Liquid chlorine and muriatic acid are generally sold in 1-gallon (4-liter) plastic bottles, four to a case. Some manufacturers provide heavy-duty, reusable bottles and plastic cases and charge a deposit to ensure return. Others supply thinner, disposable bottles in cardboard cases for a slightly higher price, but you don't have to mess with returns or the expense of lost bottles and cases.

When adding liquid chlorine or acid to a pool, pour as close as possible to the water's surface to avoid splashing it on your clothes or deck, discoloring both. Air and sun contact will diminish the concentration of chlorine, so keeping it close to the surface of the water also minimizes those impacts. Add the chlorine slowly near a return line while the circulation is running to maximize even distribution. Pour some near any known dead spots in the pool, or directly over any appearance of algae. Never pour liquid chlorine into the skimmer. Avoid skin or eye contact with chemicals.

The amounts of various products needed to raise or lower chlorine, alkalinity, hardness, and stabilizer levels are detailed in Fig. 8-4. Because of variations in products however, it is always important to read the product label to ensure the desired results.

Granulars

RATING: EASY

Sanitizers, acids, and alkaline materials are available in granular form. As with liquids, study of the product label will give you exact application methods and amounts.

Granular products tend to settle to the bottom of the pool before dissolving, so with vinyl liners it is wise to brush immediately after application. Another method is to create a solution of the product by dissolving it in a bucket of water before pouring it into the pool.

None of these products should be poured directly into the skimmer and all should be applied with the circulation running. Since granular products tend to be extremely concentrated, they can cause skin irritation or breathing problems if direct contact is made. Handle them with care.

Amount of product required to obtain a change in 1000 gallons (3785 L) of water:

Item	Raise or lower 10 ppm	Product	Amount required
Calcium hardness	Raise	Calcium chloride	2 oz. (57 g) dry weight
Total alkalinity	Raise	Bicarb of soda	2.5 oz. (71 g) dry weight
Total alkalinity	Lower	Sodium bisulfate (dry acid)	2.5 oz. (71 g) dry weight
Total alkalinity	Lower	Muriatic acid	$1/4$ cup (62 mL) liquid
Conditioner	Raise	Cyanuric acid	1.5 oz. (43 g) dry weight

Example: 20,000 gallon pool needs 30 ppm increase in total alkalinity using bicarb.
20 (thousands of gallons) $\times$ 3 (30 ppm divided by 10 ppm) $\times$ 2.5 oz. = 150 oz. required (9 pounds, 6 oz. or 4.3 kg)

Amount of chlorine needed to raise residual in 1000 gallons (3785 L) of water by 1 ppm (multiply results by 30 for superchlorination procedures):

Available chlorine of product used	Product required to raise 1 ppm
12%	$1/8$ cup (31 mL) liquid
50%	$1/4$ oz. (7 g) dry weight
80%	$1/6$ oz. (4.5 g) dry weight

Example: 20,000 gallon pool needs 2 ppm residual increase using liquid chlorine.
20 (thousands of gallons) $\times$ 2 (2 ppm) $\times$ $1/8$ cup (per ppm) = 5 cups (1250 mL)

Note:
— 16 oz. dry weight = 1 pound (454 g) dry weight
— 64 oz. liquid   = 1 gallon (3.8 L)
— 8 oz. liquid   = 1 cup (250 mL) liquid
— Use accurate measurements, such as a measuring cup or the cap of a product container after measuring the volume of the cap.

FIGURE 8-4 Chlorination, conditioner, alkalinity, and calcium hardness chart.

Granular products tend to have a longer shelf life than liquids, but time or prolonged exposure to sunlight will diminish their efficacy as well. Store and treat granular products as you would a liquid.

Tabs and Floaters

RATING: EASY

Chlorine tablets are sold to place in floating devices (sometimes shaped like ducks) which allow a slow dissolving process. Tablets are valuable when you can't service the pool for some time and need a constant source of sanitizer. They are expensive and unregulated, which is to say you can't control the rate

TRICKS OF THE TRADE: FLOATERS

- Use a floater for your chlorine tablets; never leave them in the skimmer. The low pH of tabs means you are assaulting your circulation system and any related metal parts with acid!
- To keep the floater from drifting to the skimmer opening, tie it to a return line nozzle away from the skimmer. This also makes collection easier when you need to replace the tablet.
- Don't use floaters if the pool is used by curious kids. You don't want them playing with a chemical delivery device!

of dissolution. Some floaters have a valve that allows more or less water to flow in them, theoretically controlling the amount of chemical entering the water, but the results are trial and error. Like granular products, tablets left on the bottom of a pool will bleach out any color, so use a floater.

Mechanical Delivery Devices

RATING: EASY

Santizers are the only pool chemicals generally added by a mechanical device. Mechanical delivery systems generally fall into two categories: erosion systems and pumps.

EROSION SYSTEMS

As the name implies, these systems use the water passing over the dry chemical to erode or dissolve it into solution in the water. Like the floater, the erosion system is usually controlled by restricting the volume of water allowed to pass through the system and, therefore, the amount of erosion that can take place of the tablet or granular chlorine (or bromine) inside. Figure 8-5 shows a typical erosion chlorinator system.

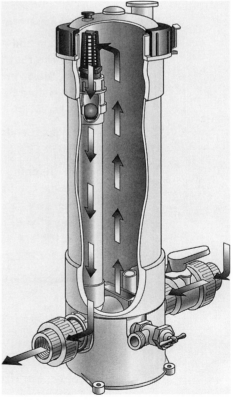

FIGURE 8-5 Typical in-line erosion-type chlorinator. *Pentair Pool Products, Inc.*

FAQS: WATER CHEMISTRY

How often must I add chemicals to the pool?

- In summer, when ultraviolet rays from the sun and heavy bather loads are depleting your sanitizer, you may need to add more (and keep pH balanced) every other day. Vinyl liners last longer if you add small amounts of chemicals frequently, rather than adding large doses all at once. In cold weather, chemical treatments may be needed only every other week.

Are there alternatives to chlorine?

- Yes. Bromine, biguanicides, and ozone are the most popular alternatives to chlorine. Some still need low concentrations (residuals) of chlorine in the water to ensure that algae doesn't get started.

What happens if I spill chlorine or acid in my pool or yard?

- Chlorine and acid together form potentially lethal chlorine gas. Be very careful when handling all pool chemicals, but especially those two. Dilute any spilled chemicals with plenty of fresh water as soon as possible to prevent killing grass or severely staining patios and decks.

Why does the algae keep coming back?

- Algae won't grow in a pool that is properly cleaned and sanitized. If you are doing both of those tasks diligently, your water may have become too hard or the pH may be too high. Keep an eye on all chemistry parameters for best results, and brush the bottom and sides of your pool often. Finally, recurring algae is often a sign of insufficient daily circulation. Increase the time of your filter run and look for improvements.

Erosion systems are typically made of PVC plastic and are located in the equipment area, plumbed directly into the circulation lines. These chlorinators are easily installed following the directions supplied, and require only basic plumbing techniques. They must always be located after any other equipment in the system. If you were to place one between the filter and heater, the concentrated chemical would corrode the internal parts of the heater.

Chemical Safety

Throughout this chapter, I have recommended that you take care when handling or dispensing chemicals or even small amounts of reagents. Here are a few additional safety tips:

1. When dealing with new chemical treatments, read all labels carefully. Apply in small amounts first, check for any unexpected results, and then continue.

2. Always take your time. Caution is the way.

3. Be careful when mixing chemicals in your pool. The table on p. 270 provides some common-sense guidelines, but always read product labels.

Pool Chemical Mixing Guidelines

	Other chlorine products	Bromine	Biguanicide
Chlorine	OK	OK	No
Bromine	OK	OK	No
Biguanicide	No	No	OK
Muriatic acid	OK	OK	OK
Soda ash	OK	OK	OK
Clarifiers	OK	OK	OK
Cyanuric acid	OK	No	No
Phosphate-based vinyl cleaners	OK	OK	No
Monopersulfate shock treatments	OK	OK	No
Algicides	OK	OK	Caution
Alkalinity/calcium hardness adjusters	OK	OK	OK
Ozone	OK	OK	OK
Ultraviolet	OK	OK	OK
Chelating agents	OK	OK	Caution

Cleaning and Servicing

Now that I have reviewed what makes a pool tick, including the components of water chemistry, let us turn to the most basic aspect of water maintenance, routine cleaning and servicing. No two technicians will approach a job in the same manner, so there is no right or wrong. You will develop a feel for what is needed week to week, season to season.

Experience and your personal style will determine how you approach a routine cleaning, a major cleanup, or other special service work. What follows is a description of what jobs you will face and what tools you will need to do the job. The rest is up to you. First I will review the tools, then describe how to use them in a typical service call and in special situations.

Tools

Figure 9-1 shows the basic service equipment carried by a professional water technician and used by do-it-yourself pool owners too.

Telepoles

The heart of the cleaning system is the telepole (telescoping pole) (Fig. 9-2A). The one you will use most on pools is 8 feet (2.4 meters) long, telescoping to 16 feet (5 meters) by pulling the inner pole out of the outer one. The end of the pole has a handgrip or a rounded tip to prevent your hand from slipping

FIGURE 9-1 Typical cleaning and service tools.

TOOLS OF THE TRADE: CLEANING

Copy and laminate this handy checklist of the basic tools you'll need to service a pool. By referring to it before leaving your garage or storage area, you'll take just the right tools each time and not make numerous trips back out for things you forgot!

- Telepole
- Vacuum head
- Vacuum hose
- Leaf rake
- Leaf vacuum
- Vinyl soap
- Vinyl brush
- Wall brush (nylon)
- Test kit
- Submersible pump and extension cord
- Drain flush bag

- Garden hose
- Multitool (Leatherman type)
- Thermometer
- Backwash hose with 2-inch (5-centimeter) hose clamp
- Notepad and pencil
- Waterproof marker
- Foam knee pads
- Sanitizer and pH adjusters
- Garbage bag

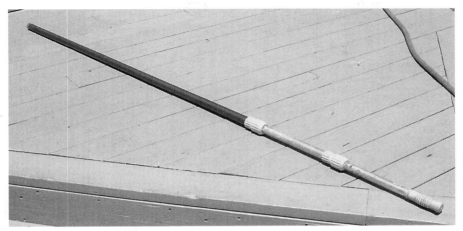

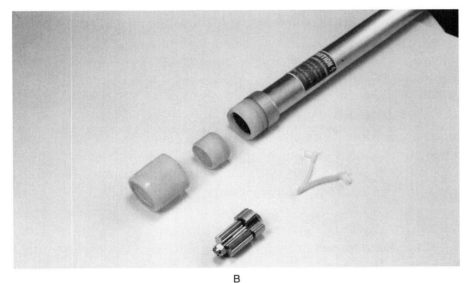

FIGURE 9-2 (A) Typical telepole. (B) Twist-grip compression nut lock, cam lock, and tool clip.

off the pole. The tip might also include a magnet for picking up hairpins or nails from the pool bottom. There are several sizes, from a 4-foot (1.2-meter) pole that telescopes to 8 feet (2.4 meters), all the way up to a 12-foot (3.6-meter) pole that telescopes to 24 feet (7 meters). To lock the two poles together, there is a cam lock or compression nut ring.

The cam lock (Fig. 9-2B) is a simple device. The cam is mounted to the end of the inside pole by drilling a hole in the center of the hub, passing a screw through the hole, and screwing this assembly into the tip of the pole. The hub is serrated, like a cog wheel, to grip the interior surfaces of the outer pole. The screw is set just off center so that it can create a circumference larger than the interior diameter of the pole. When you twist the two poles of the telepole in opposite directions, the cam swings to one side or the other in this large circumference, locking the two halves together. By twisting in the opposite direction, you unlock the poles.

When you purchase your first telepole, take it apart and observe how this cam system works. Sooner or later, scale, corrosion, or wear and tear will clog or jam the cam. Rather than buy an entirely new telepole, you can take it apart, clean it up, replace the cam if necessary, and get on with the job.

The other locking device for telepoles is a compression nut ring, like the plumbing compression rings described previously. By twisting the ring at the joint of the two poles, pressure is applied to the inner pole, locking the two together.

I have used both types of locking devices and sooner or later they both fail because of normal wear and tear. My telepole is a hybrid. I use a good quality compression ring pole, but I have disassembled it and added a cam to the inner pole. This belt and suspenders technique gives me a pole that lasts forever (seemingly) and holds up no matter what pressures I apply, particularly important when doing a cleanup of an extremely dirty, debris- and leaf-filled pool.

At the end of the outer pole you will notice two small holes drilled through each side, about 2 inches (50 millimeters) from the end and again about 6 inches (15 centimeters) higher. The various tools you will use are designed to fit the diameter of the pole. You attach them to the pole by sliding the end of the tool into the end of the pole. Small clips inside the tool have nipples that snap into place in one of these sets of holes, locking the tool in place. Other tools are designed to slip over the circumference of the pole, but they also use a clip device to secure the tool to the holes at the end of the telepole.

Telepoles are made of aluminum or fiberglass. The latter is more expensive but well worth the money. They are not only impervious to corrosion and virtually unbreakable, but they won't kill you if you inadvertently touch them to an exposed overhead wire (not that rare) or insert them in a pool with an electrical short (rare).

Leaf Rake

Figure 9-3 shows a professional, deep-net leaf rake. The net itself is made from stainless steel mesh and the frame is aluminum with a generous 16-inch-wide (40-centimeter) opening. There are numerous leaf rakes (deep net) and skimmer nets (shallow net) you can buy, but only the one pictured will last. The cheap ones are made from plastic net material and frames. Although the original price is about twice that of the cheap ones, metal ones last a long time and resist tearing when you are scooping out huge volumes of wet leaves after a windy autumn day.

The leaf rake shank fits into the telepole and clips in place as described previously. Be careful not to spill acid or other caustic chemicals on your leaf rake; both the metal and plastic mesh will deteriorate and holes will develop. Some leaf rakes are designed so you can disassemble them and replace the netting, which is fine if you have the time and patience to do it.

Wall Brush

The wall brush is designed to brush pool interior surfaces (Fig. 9-3). Made of an aluminum frame with a shank that fits the telepole, the

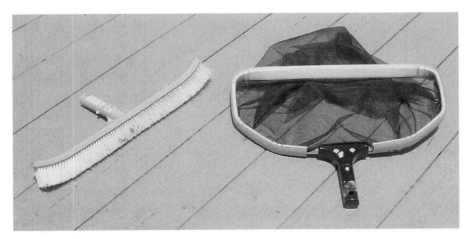

FIGURE 9-3 Brush and leaf rake.

nylon bristles are built on the brush either straight across or curved slightly at each end. The curved unit is useful for getting into pool corners and tight step areas. Wall brushes come in various sizes, the most common for pool use being 18 inches (45 centimeters) wide. Don't use stainless steel brushes in vinyl-lined pools. They are too harsh for the fabric, and the resulting scratches not only weaken the vinyl but make excellent places for algae to grow. Metal bristles will snap off and ultimately rust, leaving stains on your liner.

Vacuum Head and Hose

There are two ways to vacuum the bottom of a pool. One actually sucks dirt from the water and sends it to the filter. The other uses water pressure from a garden hose to force debris into a bag that you then remove and clean (leaf vacuum).

The vacuum head and hose are designed to operate with the pool circulation equipment. The hose is attached at one end to the bottom of the skimmer opening and at the other end to the vacuum head. The vacuum head is also attached to the telepole. With the pump running, you glide the vacuum head over the underwater surfaces, vacuuming up the dirt directly to the filter.

Vacuum heads are made of rigid plastic, with brushes on the side that touches the pool bottom (Fig. 9-4).

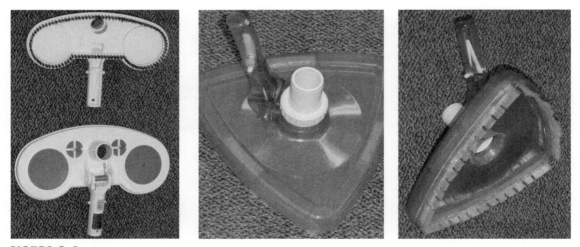

FIGURE 9-4   Typical vacuum heads (top and underside views).

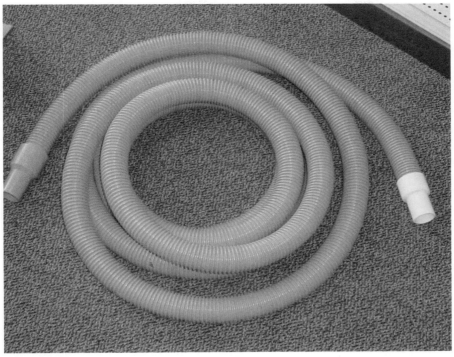

FIGURE 9-5 25-foot (7.6-meter) vacuum hose.

Hoses are available in many grades of quality, from economy models (thin plastic material) through "Cadillacs" (heavy rubber-plastic material with ribs to protect against wear), and in various lengths [10 to 50 feet (3 to 30 meters)] (Fig. 9-5). The hose cuff is made 1¼- or 1½-inch (38- or 40-millimeter) diameter to be used with similar vacuum head dimensions. Cuffs are female threaded at the end that attaches to the hose so you can screw replacement cuffs onto a hose. The best cuffs swivel on the end of the hose, so when you are vacuuming there is less tendency for the hose to coil and kink. Another valuable hose fitting is the connector. It is designed with female threads on both ends to allow joining of two hose lengths—a useful feature when you encounter a large or extremely deep pool.

Leaf Vacuum and Garden Hose

Some products become so successful in setting the standard for the industry that the brand name becomes interchangeable with the prod-

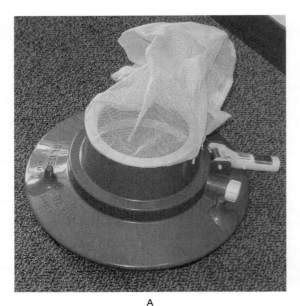

A

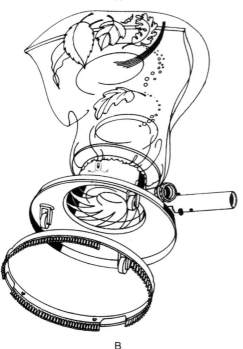

B

FIGURE 9-6 (A) Typical leaf vacuum. (B) Leafmaster with brush unit for vinyl-lined pools. *B: Image courtesy of Jandy.*

uct's descriptive name. Kleenex and Jell-O are examples of facial tissue and gelatin, but we all use the brand name regardless of the actual maker of the item. So it is with Leaf Master, the original water-powered leaf vacuum (also called a *leaf bagger*). The leafmaster, which is attached to the telepole and a garden hose, operates by forcing water from the hose into the unit where it is diverted into dozens of tiny jets that are directed upward toward a fabric bag on top of the unit (Fig. 9-6). The upwelling water creates a vacuum at the base of the plastic helmet, sucking leaves and debris into the unit and up into the bag (Fig. 9-7). Water passes through the mesh of the bag but the debris is trapped.

Fine dirt passes through the filter bag, but a fine-mesh bag is sold for these units that will capture more dirt. When the bag has a few leaves in it, they will also trap much of the sand and other fine particulate matter that would otherwise pass through.

The only other drawback to the leafmaster is if you are in a location where water pressure from the garden hose is weak. The result is weak jet action and weak suction. The other result is that as debris fills the bag, the weight of it (especially wet leaves) tips the bag over, scraping the pool floor, stirring up debris, or tangling with the hose. The latter problem is easily solved by putting a tennis ball in the bag before placing it in the pool. The tennis ball floats, keeping the bag upright.

Leafmasters are made in rigid plastic or aluminum, and for my money, the original still works the best. The wheeled version is acceptable for vinyl-lined pools, but there is also a model with brushes in place of wheels (Fig. 9-6B). One more tip: Pool supply houses sell garden hoses that float and are resistant to kinks. A hose that scrapes the pool bottom will stir up dirt and debris before you can vacuum it.

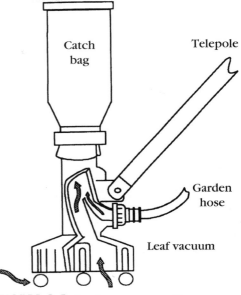

FIGURE 9-7 Leaf vacuum operation.

Tile Brush and Tile Soap

Tile brush? What above-ground pool has tile? Well, none that I've seen so far, but they all have a waterline. The same dirt, oils, and grease that foul tiles in an inground pool will darken the waterline in your above-ground model, so a tile brush and soap are handy additions to every pool tool kit.

Tile brushes are made to snap into your telepole so you can scrub the waterline without too much bending. Mounted to a simple L-shaped, two-part aluminum tube, the brush itself is about 3 by 5 inches (76 by 127 millimeters) with a fairly abrasive foam pad for effective scrubbing. If you can walk around your entire pool and easily reach over the side, try a barbecue grill cleaning pad. It has a convenient grip handle and an abrasive Brillo-type pad, which is much more effective at cleaning scum than most tile brushes.

Tile soap is sold in standard preparation at the supply house, but I recommend mixing it into another container with one part muriatic acid for every five parts soap. This will help cut the stubborn stains and oils, but it will also eat into the plastic on the tile brush pads and plastic barbecue grill brush handle, so keep rinsing them in pool water after each application and scrubbing. Don't use other types of soap in place of tile formulations, because they might foam and suds up when they enter the circulation system.

Keep your tile soap in a squirter bottle, such as a kitchen dishwashing detergent bottle. You can control the amount of soap that flows onto the brush pad and this bottle serves another valuable function. When the wind is blowing ripples on the surface of the water, you won't be able to see the bottom to know what you are vacuuming. The squirter bottle allows you to send a stream of tile soap across the pool and, as the soap spreads out over the water, it calms the ripples. You don't need much, though you might have to repeat the process two or three times in the course of a cleaning to finish your work.

Test Kit and Thermometer

Because test kits and methods were discussed previously, I will not repeat the information here. One additional word, however, is that you should keep abreast of new testing materials and kits, especially electronic testers. Most advances in this field save time, money, or both and will certainly make adding chemicals to the pool more accurate.

A thermometer is also an important item in your test kit. I keep a floating and a standard thermometer (with string). As noted in the chemistry chapter, certain tests are best performed at certain temperatures. You will also need a thermometer to check heater performance.

Handy Tools

The other basic necessities of the well-outfitted pool tool kit (as described in previous chapters) are

- Drain flush bag (blow bag)
- 100 feet (30 meters) of flexible backwash hose
- Skimmer diverters and adapters
- Spare garden hose ends (they often break and leak)
- Basic hand tools
- Tennis balls (for floating the leafmaster bag as needed and for blocking off a skimmer when the pool has more than one)
- Plastic trash bags

Pool Cleaning Procedures

Every technician, and every homeowner for that matter, approaches pool cleaning differently. After many years and hundreds of thousands of service calls, I have discovered that there are a few basic procedures that are efficient and save time.

Deck and Cover Cleaning

RATING: EASY

Most people with pools that have a deck overlook the fact that if they spend 30 or more minutes cleaning a pool, it will quickly appear as if they were never there when the leaves and dirt on the adjoining deck blow in on the first breeze. A quick sweep or hosing of any debris near the pool, at least 10 feet (3 meters) back from the edge, will keep your

---

QUICK START GUIDE: POOL CLEANING

1. **Prep**
   - Set all service tools and chemicals near the pool. Don't forget the garden hose for adding water and/or using your leaf vacuum.
   - Remove pool cover and sweep the deck.
     - Make sure pump is on and water is circulating normally.
     - Check/adjust water level.
     - Check equipment area for leaks.
     - Empty strainer pot as needed.
2. **Topside cleaning**
   - Clean the skimmer basket.
   - Skim pool water surface.
3. **Interior cleaning**
   - Vacuum the bottom with suction vacuum or leaf vacuum.
   - Clean scum from vinyl at waterline.
4. **Chemistry**
   - Check sanitizer and pH levels. Check other water parameters if scheduled.
   - Adjust levels as needed with preferred chemicals.
   - Brush bottom and sides to remove algae and distribute chemicals.
5. **Cleanup**
   - Pack away service tools and chemicals. Be especially cautious with chemicals. Be sure to keep them away from kids.
   - Replace cover as needed.

service work looking good after you have finished. Similarly, remove as much debris as possible from the pool cover before removing it. If the cover is a floating type without a roller system, be sure to fold or place it on a clean surface. Otherwise, when you put it back in place, it will drag leaves, grass, or dirt into the pool. Also be careful to avoid abrasive or sharp surfaces as you drag the cover off of the pool or you might be stuck with an expensive replacement. Finally, hose off the cover before returning it to the pool.

Water Level

RATING: EASY

If you add an inch or so of water to the pool each time you service it, you will probably keep up with normal evaporation. After rains you might need to lower the pool level. In this case, use a submersible pump and a backwash hose (or spare vacuum hose) for the discharge. Alternatively, you can run the pool circulation system and turn the valves to waste. If you use this method, don't forget to return the valves to normal circulation before you leave or you might end up with an empty pool and burned out pump.

Surface Skimming

RATING: EASY

It is much easier to remove dirt floating on the surface of the water than to remove it by any means from the bottom. Using your leaf rake and telepole, work your way around the pool, raking any floating debris off the surface. As the net fills, empty it into a trash can or plastic garbage bag (which you will notice is on your equipment list).

There is no right or wrong way to skim, but as you do, scrape the waterline, which acts as a magnet for small bits of leaves and dirt. The rubber-plastic edge gasket on the professional leaf rake will prevent scratching the vinyl. This action is often overlooked and is a rich source of small debris that will soon end up on the bottom.

One last tip. If there is scum or general dirt on the water surface, squirt a quick shot of tile soap over the length of the pool. The soap will spread the scum toward the edges of the pool, making it more concentrated and easier to skim off.

Cleaning the Waterline

RATING: EASY

The waterline of any type of pool needs special care if you want it to look good. Body oil, suntan lotion, dirt, and minerals from evaporating water will leave a "bathtub ring" on vinyl surfaces. Use your tile brush (described above) and special soap to eliminate these impurities before they become permanent stains.

Equipment Check

RATING: EASY

At this point, many people will start vacuuming the pool. Unfortunately they often get started only to discover that the suction is inadequate. It pays to check your equipment before vacuuming.

My technique is to review the circulation system by following the path of the water. Since you have already skimmed and cleaned the tile, you can now clean out the pool's skimmer basket without concern that it will fill up again while you work. Empty the contents of the skimmer basket into your trash can or garbage bag.

Next, open the pump strainer basket and clean it. If the pump has a clear lid, you might be able to observe if this step is necessary. Reprime the pump. Check the pressure of the filter. There is no point in checking it before cleaning out the skimmer and strainer baskets, because if they are full the filter pressure will be low and will come back up after cleaning the baskets. If the pressure is high, the filter might need cleaning. As mentioned in the chapter on filters, I don't believe in backwashing except as a very temporary measure. For example, if the filter is dirty and you need to vacuum the pool, you would not want to tear down and clean the filter, then vacuum dirt to it. In such cases you might want to backwash, if the filter uses that technology, vacuum, and then do the teardown.

Check the heater for leaves or debris. Look inside to make sure rodents haven't nested and to verify that the pilot light is still operating (if the unit is a standing pilot type). Turn the heater on and off a few times to make sure it is operating properly. While the heater is running, turn the pump off. The heater should shut off by itself when the pressure from the pump drops. This is an important safety check.

Of course, some customers shut down the heater in winter and don't want the pilot running, so be familiar with the heater's needs before taking action.

Check the time clock. Is the time of day correct? Is the setting for the daily filter run long enough for prevailing conditions? If the pool includes an automatic cleaning system with a booster pump, is the cleaner's time clock set to come on at least one hour after the circulation pump comes on and set to go off at least one hour before the circulation pump goes off? Always check the clocks because trippers come loose and power fluctuations or outages can play havoc with them. Service work on household items unrelated to the pool can also affect the clocks, when other technicians turn off the breakers for a period of time to do some work. Also, electromechanical time clocks are not exactly precision instruments. One might run slightly faster than another, so over a few weeks one might show a difference of an hour or more, upsetting your planned timing schedule.

At each step of the equipment check, look for leaks or other early signs of equipment failure. Clean up the equipment area itself. Remove leaves from around the motor vents and heater to prevent fires, and clear deck drains of debris that could prevent water from draining away from the equipment during rain.

Vacuuming

RATING: EASY

If the pool is not dirty or has only a light dusting of dirt, you might be able to brush the walls and bottom, skipping the vacuuming completely. If the pool is dirty, however, you have two ways to clean it: vacuuming to the filter or vacuuming with the leafmaster.

VACUUM TO FILTER

Vacuuming to the filter means the dirt is collected from the pool and sent to the circulation system's filter.

1. **Maximize Suction** Make sure the circulation system is running correctly and that all suction is concentrated at the skimmer port. If the system includes valves for diversion of suction between the main drain and the skimmer, close the main drain valve completely and turn the skimmer valve completely open. If there are two skimmers

in the pool, close off one by covering the skimmer suction port with a tennis ball, concentrating the suction in the other one. On large pools, you might have to vacuum each half separately.

2. **Set Up** Attach your vacuum head to the telepole and attach the vacuum hose to the vacuum head. Working near the skimmer, feed the head straight down into the pool with the hose following (Fig. 9-8). By slowly feeding the hose straight down, water will fill the hose and displace the air. When you have fed all the hose into the pool, you should see water at the other end (which should now be in your hand).

3. **Start Up** Keeping the hose at or near water level to avoid draining the water from it, slide the hose through the skimmer opening and into the skimmer. Insert the hose cuff into the skimmer's suction port, which may be located inside the skimmer or on the pool wall, just below the skimmer itself (Fig. 9-9). If your vacuum suction port is on the pool wall, close off the suction to the skimmer. Most skimmer lids for aboveground pools are designed to be inserted into the skimmer, slide over the skimmer basket, and thus force all suction to the pool wall's vacuum port. Always turn off the pump when moving the lid into or out of the skimmer to avoid unanticipated changes in suction that could injure your hand. If your lid is lost or not of that design, put a tennis ball over the skimmer suction port to concentrate the suction at this wall port.

FIGURE 9-8 **Vacuuming the pool.** *Hoffinger Industries/Doughboy Pools.*

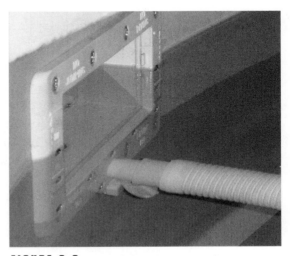

FIGURE 9-9 **Connecting the vacuum hose.**

In any case, make sure the hose does not contain a significant amount of air. When the air reaches the pump, you will lose prime. If this occurs, remove the vacuum hose, reprime the pump, then try again.

4. **Vacuum** Vacuuming a pool is no different than vacuuming your carpet. Work your way around the bottom and sides of the pool. If the pool is dirty, it will be very obvious where you have vacuumed and where you haven't. The only word of caution is to avoid moving the vacuum head too quickly. You will stir up the dirt rather than suck it into the vacuum. If the suction is so strong that it sucks the vacuum head to the pool surfaces, adjust the skimmer valves to reduce the flow.

5. **Adjust** Conversely, if the suction is weak, you might need to move the head more slowly around the pool to vacuum it thoroughly.

    If the pool is very dirty, the sediment might fill the strainer basket or exceed the capacity of the filter. You need to stop when suction becomes weak to empty the strainer basket or, in extreme cases, clean the filter.

    If the pool contains both fine dirt and leaves, the leaves will quickly clog the strainer basket. You can purchase a leaf canister, which is an in-line strainer that collects the leaves and allows fine dirt to pass on to the filter. You can empty the canister without losing prime in the pump, making your cleaning job faster.

6. **Purge** When you are finished, position yourself near the skimmer and remove the vacuum head from the water. The suction will rapidly pull the water from the hose. Be prepared to pull the hose off of the skimmer suction port before the air reaches there, otherwise you will lose prime. You can also let the vacuum run for a minute after you finish to make sure the hose contains clean water, then remove the hose from the skimmer suction port and drain the water from the hose back into the pool. Unfortunately, if any sand or small stones are hung up in the coils of the hose, they will wash back into the pool when you empty it that way, so I suggest the pump evacuation method just described. You can also pull the vacuum head from the pool and the suction end of the hose from the skimmer simultaneously, remove the hose from the water, and drain it on the deck or ground.

7. Clean Up Remove your equipment from the pool. Check the pump strainer basket and filter to see if they have become clogged with debris. Clean as needed. Replace the skimmer basket and restore normal circulation.

VACUUM TO LEAFMASTER

If the pool is littered with leaves or other heavy debris, you might need to use the leafmaster instead of the vacuum, or use the leafmaster first, then allow the fine dirt to settle and vacuum it to the filter.

1. Set Up Attach a garden hose to a water supply, then to the leafmaster. Clip the leafmaster onto the telepole.

2. Vacuum Place the leafmaster in the pool. Turn on the water supply and vacuum. As with vacuuming to the filter, leafmaster vacuuming is just a matter of covering the pool floor and walls. Because the leafmaster is large, you can move it quickly and vacuum the pool in about half the time required to filter vacuum. Be careful not to move the leafmaster so fast that you stir up the debris. If your garden hose is not the floating variety, work in a pattern that keeps the hose behind your work so you will not stir up debris before you can vacuum it.

Empty the collection bag periodically if it becomes too full (see step 3 for directions), but remember, some leaves in the bag will help trap fine dirt or sand in the bag, so leaving some debris in the bag is helpful.

3. Removal When you are finished, remove the leafmaster by turning it slightly to one side and slowly lifting it through the water to the surface. If you pull it straight up, debris will be forced out of the bag and back into the pool. Never turn the water supply off before removing the leafmaster from the pool, for the same reason. The loss of vacuum action will dump the collected debris right back into the pool. When the leafmaster is on the deck, turn off the water supply and clean out the collection bag.

Always be sure your collection bag is securely tied to the leafmaster. I can't tell you how frustrating it is to have the bag come loose as you lift the leafmaster from the pool, spilling the debris back into the pool. It can also come loose during vacuuming, especially if there is a lot of debris and strong water pressure.

Remember, if the water pressure is poor, keep a tennis ball in the collection bag to keep it upright.

Chemical Testing and Application

RATING: EASY

Some technicians prefer to test and add chemicals as needed before cleaning because the actions related to vacuuming the pool will help mix the chemicals throughout the water. Remember that dirt will destroy sanitizers, so why add costly chemicals, only to have them instantly neutralized? Test the water and add chemicals after cleaning the pool, and be sure they are evenly distributed by brushing the pool afterward or by adding chemicals near the return line when the circulation system is operating.

1. **Test** Follow the general testing guidelines described previously, testing for chlorine residual, pH, total alkalinity, and acid (or base) demand. Unless you detect a problem, these are the basic tests you will conduct at each service. Once a month, or if you find unusual readings in the basic four tests, you will test for calcium hardness and stabilizer.

2. **Treat** Apply the chemicals as described previously. I use liquid chemicals for ease of distribution and price, but you will make your own judgments as you proceed. In any case, it is purely a personal preference. Apply sanitizer, following the guidelines described previously, then move ahead to the brushing step to give it time to circulate and distribute. Brushing accelerates that distribution. Apply acid or alkaline materials next. You don't want to apply them together because combinations of pool chemicals can be deadly.

3. **Caution** Be careful with chemical bottles on pool decks or grass lawns. Like the ubiquitous coffee cup ring, pool chemical bottles will leave stains on decks and kill grass in patches that are difficult to remove and even more difficult to explain. If you must set them on the deck or grass, immerse the bottom of the bottle in the pool for a moment to rinse off any chemical. Splash some water on the deck where you set the bottle down so any residue will be neutralized and not leave stains.

Brushing

RATING: EASY

Having now spent 30 to 45 minutes servicing the pool, many pool owners skip the brushing. This is a big mistake. Dirt and minerals will build up over time, making the pool appear dingy and dirty even when you have cleaned it thoroughly. Also, algae starts out invisible to the naked eye, and brushing removes it from surfaces, suspending it in the water where chemicals will kill it before it can really take hold.

On pools that are not very dirty, you can skip vacuuming and brush from the walls to the bottom and from the shallow to the deep end, directing the dirt toward the main drain (if your pool has one) where it

---

TOP 10 THINGS YOUR POOL CAN DO FOR OUR ENVIRONMENT

1. Buy products that come in reusable or recyclable containers.

2. Take out-of-date chemicals and cleaning supplies to a hazardous waste recycling center.

3. Add a cover or solar heating system instead of using gas heat.

4. Don't run the pump more hours per day than is needed to keep the pool clean.

5. Save water by keeping the cover on when the pool is not in use and by asking swimmers to avoid excess splashing of water out of the pool.

6. Save water by checking for leaks every week (and repairing them promptly) and by using a broom instead of a hose to clean pool decks.

7. Keep leaves and other organic debris out of the pool—they destroy sanitizers and other pool chemicals, requiring you to buy and add more.

8. Replace the net on your leaf rake instead of buying a new unit. Other service tools also have replaceable parts, so ask before you throw usable tools into the landfill.

9. Use your pool on hot days to cool off rather than using the air conditioner.

10. Buy an energy-efficient motor when replacing one on your pump.

is sucked to the filter. If you plan to use this technique instead of vacuuming, divert all suction to the main drain as described previously.

Cleanup and Closeup

RATING: EASY

After you have collected all your belongings, take one last look at the pool before leaving. Did you add enough water and turn off the water supply? Did you pick up everything you brought with you?

Service Schedule

So just how often should you perform the various service and maintenance tasks discussed in this book so far? The following chart offers some guidelines, but frequency of service items depends on conditions at your pool, such as how often it is used, weather factors, and how dirty it gets. Experience is the best teacher.

Periodic Service and Maintenance Schedule*

	Day	Week	Month	Year
Check water level.	1X			
Check pH and sanitizer.		2X		
Check hardness, TDS, and total alkalinity.			1X	
Test for heavy metals.				2X
Check conditioner.				2X
Check skimmer basket.		2X		
Check pump strainer basket.		1X		
Check filter pressure.		1X		
Look for leaks in plumbing and equipment; check time clock settings.		1X		
Surface skim the pool.		1X		
Vacuum the pool (pool has no auto cleaner).		2X		
Vacuum the pool (pool has auto cleaner).		1X		
Operate auto cleaner (3 hours).		2X		
Brush pool walls and bottom.		1X		
Clean waterline.		1X		
Empty auto cleaner catch bag.		2X		
Clean solar panels.			1X	
Winterize.				1X
Reopening and/or equipment tuneup.				1X
Tear down and clean filter.				3X

*Adjust as needed during months of heaviest use or extreme weather. ✕ = times per.

TRICKS OF THE TRADE: VINYL LINER CARE

- Low pH is common in vinyl-lined pools because of heavy bather loads and no plaster to neutralize the acidity. Acidic water can cause wrinkling and premature failure of vinyl liners, so pay special attention to pH testing and adjustments.

- High chlorine concentrations (above 1.5 ppm) can weaken vinyl, so test and adjust sanitizer levels frequently rather than adding heavy doses at one time.

- Never let granular or tablet sanitizers come into direct contact with vinyl liners. Dissolve these products in a bucket of water before adding them to the pool, or use a floater.

- "Bathtub ring" grime actually weakens the liner material. Keep the waterline clean with vinyl-safe cleansers. Apply vinyl "sunblock"—special nonpetroleum liquid (available at pool supply stores) that blocks the harmful impact of ultraviolet rays—to exposed areas of the liner.

- Never lower the water level more than half of the pool capacity. Liners will shrink and wrinkle, and they may tear at stress points when they are refilled.

- Use only brush-bottom vacuums and automatic cleaners (not wheel versions).

FAQS: CLEANING AND SERVICING

Should I hire a service or clean the pool myself?

- Most routine pool maintenance tasks are easily performed by family members, but weekly cleaning service is not expensive and can save money in the long run. Equipment repairs, annual tuneups, and replacements or additions are probably best left to the pros. Many people enjoy cleaning their own pool (it is cheaper than therapy!), so try it before deciding to hire a service.

How often should I vacuum the pool?

- At least once a week, unless you have an automatic cleaner that is working extremely well. Remember that vacuuming will simultaneously brush the pool, so it will not only keep the pool clean but also prevent algae from getting started.

How do I get rid of the "bathtub ring"?

- Discoloration around the waterline of the pool is normal, and it is caused by the minerals in water that has evaporated. Suntan lotion and body oil will cling to the waterline too, actually eating into the vinyl if not promptly removed. Use a vinyl cleaner approved for above-ground pools, and keep the waterline as clean as possible for aesthetics and to extend the life of your liner.

Repairing and Winterizing Above-Ground Pools

Now that you know how to perform the basic cleaning, chemical application, and repair services already outlined in this book, there are several additional procedures that are easy to learn and can save expensive service calls.

Draining a Pool

RATING: EASY

Repairing a liner or preparing it for winter may involve draining your pool. It might seem obvious that you use a submersible pump to drain your pool, but this simple task can create many problems if you don't take all factors into consideration.

1. **Shut Down** Turn off all circulation equipment at the circuit breakers, so there is no chance it will start up from a time clock. Be sure that underwater lights are also switched off and not connected to a time clock. Often there are switches inside the home: tape over these with a note to make sure all family members know there is a reason to keep that switch off.

2. **Safety** If your pool has a deck, run yellow caution tape around its perimeter to keep unwary visitors from falling into an empty pool. Turn deck furniture on its side and use it as a physical barrier as well.

3. **Drain the Pool** When draining a pool with your submersible pump and hoses, direct the flow to a patio deck drain if possible. This will send the waste water through an intended channel, rather than over a backyard garden or down a hill where erosion damage can occur from such a large volume of fast-moving water. Once you have started pumping, watch the flow into the drain for several minutes. Sometimes debris will back up in the line and it will overflow, but not until it has filled several hundred yards of pipe. The extra time it takes to make sure the water is flowing properly is well worth it because a clogged drain can flood backyards and living rooms, and it might actually flush the water back into the pool.

If deck drains cannot accommodate the flow, connect several vacuum or backwash hoses together and run the waste water into the street where it is carried to storm drains. Again, watch the flow and make sure that it does not back up the storm drain and flood the street.

Since the equipment for most above-ground pools is located below the waterline, you can also simply allow gravity to drain the pool. Figure 10-1 shows a three-port valve that is plumbed to send the water to the pump or to an open pipe for draining the pool. Set the valve to the drain line position, and you can easily empty the pool down to the level of the plumbing, although this may not finish the job if the equipment is located above the actual bottom of the pool.

If the three-port valve is plumbed between the pump and filter, you can turn the valve to the drain position and allow it to drain the pool, or speed up the process by actually using the pump. If you are using the pump to drain the pool, set the three-port valve on the suction side completely to the main drain, and be sure to turn if off when the job is done so the pump doesn't run dry. If you don't have a main drain on your pool, this method won't work since the pump will starve for water when the level drops below the skimmer.

Either way, securely attach a backwash or vacuum hose to the end of the drain pipe to divert the water to a deck drain or other appropriate disposal location as described above. Let me repeat that part about "securely" fastening the hose because if it comes loose, you could flood the equipment, including the electrical components of the motor. Use at least two hose clamps, and then watch the draining process for several minutes to be sure it won't come loose.

FIGURE 10-1 Pool drain line at the pump.

Smaller above-ground pools that have no main drain or skimmer can best be drained by disconnecting the suction hose, either at the pool or the pump, and allowing it to drain that way. Never open the strainer basket of the pump to drain the pool—you might flood the motor. Finally, some smaller pools have a drain plug, usually located near other plumbing fittings for ease of identification.

In areas where water conservation is a concern, you should wait to drain the pool until the chlorine residual has gone below 1 ppm, then let the flow irrigate lawns and gardens. Obviously a 20,000-gallon (75,700-liter) pool will flood the typical backyard garden if it is pumped all at once, but use as much as you can before switching the discharge to a regular drain where the water is wasted. Another technique along these lines is to punch holes along the length of a backwash hose (or old vacuum hose) and seal up the discharge end (tie a knot in a backwash hose or use plumbing fittings to close a vacuum hose). This then acts as a huge sprinkler, evenly distributing the water all along the length in various

---

TRICKS OF THE TRADE: SUBMERSIBLE SAFETY

- When lowering a submersible pump into a pool, never do so by the cord. Attach a nylon rope to the bracket or handle and lower it that way to avoid pulling electrical wires loose. Not only will the pump fail if the wires come loose, but when you plug in the cord it might electrify the water.

- Buy a ground fault interrupter (GFI) that can be plugged into the wall socket before plugging in the cord of the pump. You can also rig a remote on/off switch that plugs into the socket as well, allowing you to operate the pump while you are working in the pool. Check your local electronics store for a combination remote receiver/GFI or simply buy one of each and plug them in together before plugging your pump into the outlet. It will be the best $30 to $50 you ever spent. I have been in water with a pump that had a slight short and felt the tingling of the electricity conducted through water and have known technicians who have been killed by not taking this aspect of operational safety seriously.

---

gardens, perhaps even across the lawns of several neighbors (with their permission, of course). In other words, be creative about your method of pumping and the potential uses of the waste water.

One other word of warning about deck drains. They are usually made of PVC, but because they don't carry water under pressure they are not usually pressure tested. If ground movement or other erosion has destabilized them, the pipe might have separated and much of your waste water will end up eroding the ground under your patio. If you have any doubts about the integrity of the deck drain or its ability to handle the water, run the hoses into the street.

In some jurisdictions there might be restrictions on pumping out pools relative to the permissible volume and even the permissible chemical makeup. Extremely low-pH water might have to be neutralized before pumping it into municipal stormwater or sewer lines. Check your local codes before turning on the pump.

Leak Repair

As previous chapters have detailed, there are many ways that a pool can leak through the equipment and plumbing. There are an equal number of ways that the vessels themselves can leak.

When you have an unexplained loss of water, it makes sense to look for the obvious first. Signs of water in the equipment area and exposed plumbing will usually yield results; however, buried plumbing commonly leaks where it can't be seen. While the worst leaks require excavation of plumbing or rebuilding of the pool itself, the most common leaks can be repaired effectively. In this section, I will examine how to find those more common leaks and discuss their repair techniques.

Leak Detection Made Easy: Four Tests

As noted previously, the first place to look for leaks is in the exposed equipment and plumbing, but there are often visible signs of leaks that are otherwise hidden from view. Wrinkles in the pool liner might be a sign of leaks. Uneven framework may suggest structural leaks from shifting ground (which might have been caused by a water leak eroding the soil). Tree roots might lift the plumbing or pool wall as well. If, however, there are no such visible signs, you need to look elsewhere. Some leak detection methods will help to both verify a suspected, visible leak or locate a hidden leak.

EVAPORATION TEST

RATING: EASY

Before elaborate leak testing, you will want to verify the water loss. The most simple method is to fill a bucket and place it on the deck or ground next to the pool. Mark the level in the bucket with an indelible felt-tip marker and do likewise for the water level in the pool. Turn off the circulation to eliminate any variables in evaporation. After several days, mark the new level of water in the bucket and pool. They should evaporate an equal number of inches (or centimeters). If the pool level lowers significantly more, a leak is likely. If both vessels lower a similar number of inches, then there is no leak.

If the water in the pool drops faster than the water in the bucket, continue the test until the rate of evaporation is the same in both vessels. This process may take several days, but when the pool no longer evaporates faster, you have found the level of the leak. At this point, it will be easier to use the dye test to find the exact location.

DYE TESTING

RATING: ADVANCED

Perhaps the simplest method of detecting a leak in the vessel itself is to shoot a dye test. As the name implies, a colored dye is disbursed in suspected areas, and, as the dye disappears, the leak is found.

1. **Prep** Clean and brush the pool thoroughly. Minor holes in vinyl can sometimes be hidden by dirt or other material settled in the crevices. Pay careful attention to corners and around fittings.

   Turn the circulation off and begin the examination on a calm day. Wind rippling the surface makes it difficult to see small tears. You might also want to squirt a little liquid soap across the surface to sharpen visibility further (and you might need to repeat the process from time to time during the exam).

2. **Inspect** Examine the pool for wrinkles and tears, beginning with the waterline.

   If you find probable leak locations along the waterline, don't just stop there. If tree roots, shifting ground, earthquakes, erosion, or other problems have caused leaks at one point, they might also exist elsewhere.

   Therefore, continue to visually inspect the interior surfaces, looking for abrasions, tears, or discolored patches of vinyl, noting anything you suspect.

3. **Shoot Dye** The dye test can easily be conducted using an old test kit reagent bottle or similar squeeze bottle filled with food dye (available at any grocery store). Nontoxic dyes are sold for this purpose in pool supply stores, but food dye is just as good and less costly. Some technicians use phenol red, but you will need to check many locations, and it is unwise to inject that much acidic chemical into the water.

   Work around the pool, particularly in the locations of suspected leaks. You will need to get into the pool, including diving to the main drain, to do a thorough job. At each wrinkle or suspected area, aim the nozzle of the bottle at the wrinkle. Squeeze a bit of dye into the area and watch it. If the dye simply swirls around the wrinkle without being sucked in, then there is no leak in that area. If the dye is sucked in, it is riding on a flow of water leaking from the pool. The speed with which the dye disappears will help you estimate the size of the leak.

4. **Observe** As with the visual inspection, continue around the entire pool looking for leaks to exhaust all possibilities. Be especially careful around skimmers or other fittings. Don't forget the interior of the skimmer and the main drain, as well as the return outlets.

DRAIN-DOWN TEST

RATING: EASY

If you have tried the evaporation test and dye testing, or if you are not a good diver and wish to skip the dye test, try the drain-down method.

1. **Prep** Turn off the equipment and mark the level of the water in the pool.

2. **Mark** Mark the level again at the same time each day to establish a rate of leak. Because of normal evaporation, the level will continue to decrease indefinitely; however, the objective is to determine when the level stops lowering as a result of the leak. If, for example, you record a loss of 2 inches per day for four days, then the rate slows to 1 inch every five days, you will know the level at which the rate slowed was the level of the leak. Mark that level of rate transition.

3. **Inspect** Examine all possible leak areas along the transition level. The leak must be along this line. For example, if the water loss slowed when it reached the level of a particular return outlet, you might reasonably suspect the leak in that plumbing line.

   The only fault with this method is that it is an indicator, not a precise tool. Since water seeks the same level in all plumbing and parts of the vessel, the leak might stop at the level of a certain plumbing fixture, but actually be in an entirely different location that is coincidentally at the same level. In any case, the new level will tell you where to look further and where you need not look.

LEAK DETECTORS AND PRESSURE TESTING

RATING: PRO

When the previous methods fail to help you locate the leak, especially if you have buried plumbing, there are two other methods of leak location. There are electronic listening devices called *geophones* that can actually hear water dripping or flowing. By applying such devices around the pool and related plumbing, an operator can identify where water is moving out of the system. Because these devices are expensive and their operation requires a great deal of experience and skill, most service technicians don't buy or use them. There are numerous professionals who do, however, and they are easily found in the phone book or through referrals at your supply house.

The second method used to find leaks is pressure testing equipment. It is not difficult to pressure test a plumbing system, but the amount of time and additional equipment (plugs, adapter fittings, compressed air, and related fittings) makes this type of testing impractical for most pool technicians. Many pool builders and plumbing contractors are equipped to pressure test pool systems. Companies that conduct leak testing might also conduct pressure testing.

Repairing the Leak

RATING: ADVANCED

There are numerous manufacturers of vinyl liners and soft-sided materials for above-ground pools, using several different compositions of vinyl, PVC, or rubber-based materials. Because of the different qualities of each make of pool or liner, there are no standard procedures for repair or replacement. Having said that, however, working with such fabrics isn't difficult because they are sold with detailed instructions and repair kits.

Most repair kits detail the process of drying and cleaning the area to be patched with a solvent, roughing up the area to be glued with sandpaper for a better bond, then gluing a patch of the same material over the tear or puncture. These are simple procedures that you have used since childhood on various repairs around the home, and they should present no unusual problem on your pool.

The only caveat is to be sure to use the right kit for your pool or

TRICKS OF THE TRADE: REPLACING A VINYL LINER

Vinyl liner replacements will come with directions, but preparation is the key. Essentially, you will remove the old liner (after removing fixtures first and then disengaging it from the track and bead receivers), measure the pool, order the replacement liner, stretch it out into the pool, secure it at the rim and pool suction/return fittings, pull out the wrinkles with suction from a vacuum cleaner, and fill it with water. Here are some tips to make the job easier and more successful:

- Measure, measure, measure! The more details you can provide, the more tailored the replacement liner will be (Fig. 10-2). Be sure to use a steel tape measure for accuracy. Photos and written descriptions help the manufacturer get the subtleties of your project.

- Make careful notes of gaskets needed when replacing the fittings, and then order new ones.

- Use the opportunity of the empty pool to inspect the walls for damage.

- Level the ground before installing the new liner, and look for any points of abrasion.

- Never drag the liner when installing it—abrasions will lead to leaks later if not immediately. Also, if the subsurface is sand, dragging the liner will cause the sand to pile up at one side and create depressions elsewhere.

- The liner will be somewhat smaller than the pool to allow for stretch when filled with water, so don't panic when it doesn't seem to fit at first.

- Don't install a new liner on a very hot or cold day. The heat will cause the material to stretch too much and you'll never get the wrinkles out. The cold will prevent adequate stretch when filling to fill out the entire pool. Ideal working temperatures are between 50° and 80°F (or 10° and 27°C).

- Start the installation at the shallow end because it is easier to stand on the floor of the pool while attaching the liner to the wall.

- The manufacturer will mark the liner clearly, so start at an obvious place of alignment such as a corner or skimmer.

- Set down water bags (or buckets of water placed on towels to avoid making permanent impressions in the liner) as you lay out the liner to secure it as you work.

- Keep the vacuum running when filling the pool, at least until it is filled to within 8 inches (20 centimeters) of the surface. That will prevent wrinkles.

- On deep ends where a ladder is needed to get in or out, set the ladder on a board that is covered with a rug to avoid liner damage.

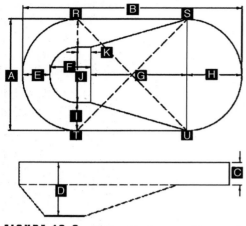

FIGURE 10-2 Measuring for a replacement liner.

liner. Because the composition of vinyl materials differs widely, glue and patch material designed for one liner might not adhere to another.

Another labor saver that is available for aboveground pools of all types is the underwater patch kit. Some kits are sold as self-adhesive patches, and others require the application of a liquid adhesive that sets up to a gumlike consistency before you apply the patch to the fabric under water. Both types of patch kits work well on smooth, flat surfaces, but they are not designed to fill gaps in through-wall fixtures and plumbing. There are liquid sealers and caulks that can be applied under water (or to wet surfaces) that will be sucked into the hole or gap as the water is leaking out. Once in the opening, these materials set up and seal the leak, but I have found these to be only temporarily useful. Typically, the same stresses that caused the leak in the first place will weaken these stop-gap measures and the leak will recur.

There is a "one-size-fits-all" repair adhesive sold by FeherGuard, makers of pool covers and roller systems, called *Pool Fix*. Unlike other glues, it does not work by exposure to air, but rather in the absence of it. Therefore, you can apply some of this glue, but nothing will happen until you press the two sides of the patch together and force all of the oxygen out. It works on all types of flexible and rigid materials, and, because no solvent is involved, it is safe for all above-ground pool uses.

If the leak is severe or has recurred in the same places, it might be time to replace the liner. Follow the steps in "Typical Above-Ground (Rigid-Sided) Pool Installation" and the professional tips in "Tricks of the Trade: Replacing a Vinyl Liner," both in Chap. 1. Here are a few additional tips about selecting the right replacement liner:

■ Avoid textured or embossed liners. The crevices catch dirt, become weak spots in the liner, and are breeding grounds for algae. If you do buy an embossed liner, be sure to ask the minimum thickness of the finished material: Is the nominal thickness measured at the high spots or the indentations in the embossing?

- If your liner is still under warranty, compare the cost of buying a new liner against the pro-rated cost of obtaining a replacement under that warranty. Many liner manufacturers offer 10- or 20-year warranties on liners knowing that the likely life of the product is not more than 5 or 7 years. They pro-rate the cost of the replacement liner based on the number of years left on the warranty, but this can often be more than the cost of buying a new liner of equal quality from a discount pool supply store (or online merchant).

- If your pool uses an overlap liner, consider upgrading to a J-bead, V-bead, or unibead version. You'll have the minor added expense of the liner hanger that must be added to the top edge of the pool wall, but you might prefer the tailored look of these styles and the ease of installation.

- Prevent liner leaks in the first place by covering bolt heads or other potentially abrasive hardware with duct tape or clear nail polish. The nail polish also prevents hardware from rusting.

Winterizing

If you live in a cold climate, I probably don't need to tell you that a body of water needs to be prepared for the winter months. But even in warmer climates, you might discover that some of the following information is useful.

The most important single concern from cold temperatures is freezing. Water expands when it freezes, meaning that if it is trapped inside pipes and equipment it will expand and crack them. PVC pipes, fittings, and ABS plastic pump parts are most susceptible, but soft copper heat exchangers and galvanized pipes are not exempt either. Expanding frozen water will also crack plastic skimmers.

The second problem that relates to the seasonal winter closure of a pool is the potential damage from algae and debris. Although algae do not grow well in temperatures below 55°F (13°C), especially if the water is shielded from sunlight by a cover, prolonged periods of stagnation will permit and promote algae development.

Of course, the easiest way to winterize an onground pool is to drain it completely and, if possible, put the entire pool and equipment in storage for the cold months. This is not practical for above-ground pools, however, because it is time-consuming, stressful on pool liners and plumbing, and wastes water, so you probably need to prepare the pool for winter and leave everything in place. One last note of caution: Even if you drain your pool, don't think the job is done. Water left in pipes and equipment can cause damage to plumbing and expensive components, so read through this section carefully and follow those procedures that apply to your pool.

Temperate Climates

RATING: EASY

For normally temperate climates, the best protection during winter is to keep an eye on weather forecasts. When temperatures are predicted to drop near freezing, set the circulation to run 24 hours per day, preferably with the heater on at the lowest thermostat setting. This is especially important for pools in valley or mountain regions where temperatures might be slightly colder than those predicted in metropolitan areas.

I find prevention to be easier than reaction. During winter months, I set my customers' time clocks to run 2 hours during the day, 2 hours in the late evening, and 2 hours in the early morning. This way, the water circulates and the heater comes on every few hours and prevents any freezing from setting in. I usually give customers an idea of the cost of replacing equipment and digging up broken pipes versus a few dollars over the winter for the heater. I haven't had a single refusal.

Continue normal service during the winter. Wind and rain means that pools will flood, and stain-causing leaves and debris can harm the vinyl surfaces or clog the system. In short, the work in winter doesn't lessen; it simply changes slightly. You will use less chlorine to maintain the same residual, and you can reduce filtration times to a total of 4 hours per day (staggered as described earlier) Of course, if the pool is kept heated, it's business as usual.

TOOLS OF THE TRADE: WINTERIZING

- **Submersible pump**
- **Pool-safe liquid antifreeze**
- **Tapered rubber winterizing plugs (one for each return outlet in the pool)**
- **Air pillows**
- **Plastic jug with sand or pebbles (that fits inside skimmer)**
- **Chemical test kit**
- **Sanitizer**
- **Tape (to obstruct electrical switches and breakers when shut off)**
- **Compressed air, hose, blow bag**
- **Pliers to remove drain plugs from pump, heater, solar panels**
- **Any hand tools needed to remove your equipment for storage**
- **Winter cover (and clips, cords, or weights to secure it)**
- **Yellow caution tape**

Colder Climates

RATING: ADVANCED

In climates where winter means actually wearing long pants and jackets, most pools are closed for the season. The shutdown process requires some planning to avoid damage to the installation and to make reopening in the spring a simple task.

1. **Balance the Chemistry** Etching or scaling conditions of water will harm the pool even when the circulation is off, so before closing make sure the chemistry of the water is balanced.

2. **Cleaning** Dirt and debris left in the water during long periods of stagnation will leave stains and/or be much harder to remove several months later. Therefore, thoroughly service and clean the pool before closure.

QUICK START GUIDE: WINTERIZING

1. Prep
   - Bring all tools and supplies to the pool area.
   - Clean and service the pool normally before lowering the water level. Increase sanitizer level per package directions.
   - Shut off
     - Pump (tape over the switch or breaker so that no one can turn it on before spring).
     - Electrical source to heater electronic ignition.
     - Gas supply to heater.
2. Drain
   - Lower the water level below skimmer and return outlet.
   - Clear lines of water and plug them.
   - Fully drain all equipment, even if it is to be stored indoors (unless stored in heated location). Remove and store equipment (if possible).
3. Secure the pool
   - Install airbags (including small airbag or plastic jug in skimmer).
   - Install winter cover.
   - Block off pool entry with caution tape or other barriers.
   - Pack up and store all service tools, especially chemicals.
4. Reopen in spring
   - Perform the same tasks in reverse order.

3. **Algae and Stain Prevention** To prevent algae growth and staining during the closure, superchlorinate the water. A simple formula is to triple the normal superchlorination that you use for the particular installation. A more precise approach is to raise the residual to 10 ppm. Some technicians recommend lower residuals for vessels that will be covered or those that will be closed for less than 4 months; however, I would stay with these recommendations regardless. Also, add a chelating agent to prevent metals from dropping out of solution and staining the surfaces. Finally, add an algicide that will inhibit black algae growth (I prefer silver-based products).

Do not leave tablets or floaters in a pool during closure. Since the water isn't circulating, the dissolving chemical isn't either, and the extreme concentrations in one area can do structural or cosmetic damage. A vinyl pool owner told me he left a floater in his pool during one winter and found it had sunk when he reopened the pool in the spring. The concentration of chlorine on that spot not only bleached a large area of the blue vinyl, it weakened the area and caused leaks later in the summer that required extensive patching.

Make sure that you have adequately circulated the chemicals before shutting off the system for the winter. Finally, there are several winterizing kits on the market with chemicals, antifreeze, and instructions. If you are new to winterizing, try one of these the first time. Even for old hands, you might discover that a package deal is less costly than buying supplies individually, because there are often specials offered by chemical companies.

4. **Shut Down the Equipment** Turn off the circulation equipment at the breakers and tape over them to prevent someone from turning them back on. Turn off all manual switches and time clocks and remove the trippers as an added precaution in case someone does turn the breakers back on. Some technicians don't disconnect underwater lights if the water level will remain above the fixture all winter. The light provides a little warmth to the water periodically and draws the owner's attention to the pool for regular inspection. I have seen light lenses crack, however, as the extremely cold water and extremely hot light fixtures contact, so I don't recommend this practice.

5. **Lower the Water Level** Pump out 24 to 36 inches (60 to 90 centimeters) of water, or at least enough to drop the level 18 inches (45 centimeters) below the skimmer line. When water freezes and expands, it can crack the plastic skimmer components, so the goal is to lower the level to a point where winter rain or snow will not raise it back up to those delicate areas. If the equipment at the installation permits, lower the level by vacuuming to the waste line on your final cleanup. This will save you time by accomplishing two goals at once.

If possible, the vessel should not be drained completely for the season because shifting ground might cause ripples when empty that will not smooth out completely when refilled. These folds in the vinyl can then stretch and weaken the material. Complete draining is not advisable unless you are also prepared to spend a great deal of time carefully smoothing and refilling in the spring.

6. **Clear the Lines** Perhaps the most important objective of winterizing is to protect the plumbing from freeze damage. There are several methods, some easier than others. The method you choose might depend on your equipment, skills, and the availability of water in your area.

The most effective method of protection is to drain the pool, completely evacuating the lines. Of course, you will then refill it to the level discussed in step 5. This temporary draining of the entire pool (not necessary if you don't have a main drain) won't harm the liner because it won't have time to shrink before you refill it to the winter level.

If water needs draining anyway for chemistry reasons, as discussed in the chemistry chapter, this is an opportunity to accomplish two tasks at once. The cost or availability of water in your area might prohibit annual draining, but if you can, it is the most effective way to be sure all water has been removed from the system. Even small amounts of frozen water can crack pipes and fittings.

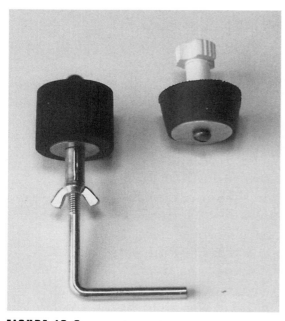

FIGURE 10-3 Test plug and tapered winterizing plug.

With the pool empty, remove the collar/nozzle fittings from the return lines and plug them with tapered expandable rubber winterizing plugs (Fig. 10-3). Be sure to use tapered plugs because they will pop out if freezing water in the pipe expands. You're better off with a popped plug than a cracked pipe. The main drain might still have water in the bottom of the plumbing, so try to suck it out with a wet/dry shop vacuum or mop it out with a sponge. Just to be sure, pour a cup of antifreeze into the main drain before plugging it. For antifreeze, use a mixture of one part propylene glycol to two parts water. Your supply house will have propylene glycol or premixed products. Never use automotive antifreeze, which is corrosive. With all of the lines plugged, refill the pool to the level described previously.

If you are unable to drain the pool, you can blow the lines out with air (described later) and plug them underwater. Some references suggest you can plug the lines first and then use a vacuum to suck water from the lines, but it is unlikely that all of the water (or even most of it) will be removed. If you have no way to blow the lines, you might try using a wet/dry vacuum, preferably a high-powered model available at tool rental shops. After you have vacuumed as much water out of the lines as possible, pour the antifreeze solution into the pipes by pouring it into the strainer pot and allowing it to flow back toward the pool. If there is a check valve in the line or if the pump sits below the water level of the pool, the mixture will not get any closer to the pool than the valve. If that part of the plumbing is exposed or you can access the suction lines by disassembly of the three-port valves or gate valves, pour the mixture in at those points.

The return lines can also be filled with antifreeze by starting the pump with the strainer pot lid open. With the pump running, pour the mixture into the pump. It will be distributed throughout the equipment, plumbing, and return lines. When you see antifreeze discharging from all of the return lines, turn the system off and plug the return outlets. You might want to add some food dye to the mix to make it easier to see when it is discharging to the pool. Some water will remain in the lines, so as you can see, without draining the pool, clearing the lines is guesswork at best.

The other method of clearing lines underwater is to blow them with air. You can buy or rent a tank of compressed air (useful for clearing obstructions from lines also) from a plumbing supply house or a company that provides compressed gasses [low-pressure tanks or blowers only—not more than 5 pounds per square inch (344 millibars)]. Attach a reinforced rubber line with a garden hose attachment on the end and add your blow bag to the end. Insert the blow bag into the suction side of the pump (be sure to get on the pool side of any check valve and remember to open any gate valves) and force compressed air through the lines. Go to the pool while the air is bubbling out of the lines and plug them. The lines will be filled mostly with air, but some water remaining in the lines is inevitable with this method. Do the same on the return lines, inserting the blow bag at the heater outlet pipe or disassembling a valve on the return side of the equipment system.

The blow-out method requires diving on the main drain to plug it. If you are unable to do that, you can try to flood the area with antifreeze. Insert a long 2-inch PVC pipe into the pool, placing one end on top of the main drain. Pour the antifreeze mixture into the pipe. Since the mixture is heavier than water, it will sink to the bottom and flow into the main drain plumbing.

One last point about draining water out of the pool circulation system: Several companies make special covers that bolt over the skimmer to seal it off from the pool water in winter. The notion is that by isolating the skimmer, you can drain it and the plumbing without lowering the pool level and wasting the water. It's a good idea, but not a practical one. First, a summer of heavy use probably warrants replacement of some of the pool water anyway (see Chap. 8). Second, these skimmer covers require removal of the skimmer faceplate and reinstallation of it in spring. Although this gives you a chance to replace the skimmer faceplate gasket, it is likely to cause leaks. Other versions snap over the skimmer faceplate like the lid of a Tupperware container, but in practice, they leak. If they fail, the water that leaks into the skimmer can freeze and cause cracks. In summary, I don't recommend these skimmer covers as a substitute for lowering the pool level.

7. **Remove Equipment** Any equipment in the pool or circulation system that might be damaged by exposure to the elements (or which might be stolen) should be removed and stored. If the pump and motor are plumbed with unions, you can easily remove the entire unit and disconnect the electrical connections. If there are no unions, you might want to cut the plumbing and add unions when you reinstall the unit in the spring to make the process easier next season. If the plumbing makes removal of the pump difficult, unbolt the wet end and motor from the volute and remove those.

The filter should be disassembled, cleaned, and drained thoroughly and the grids or cartridges put in storage. Freezing water can cause fabric deterioration. Sand filters should be cleaned and drained. Since rain might refill an open filter tank, close the tank and leave the drain plug out.

Close the gas valve to the heater and any supply valve (if it is a dedicated line for the pool heater) at the meter. The heater has drain plugs on both sides. I recommend removing the heat exchanger and burner tray and storing them after draining. Drain the water from the pressure switch tube as well (see Chap. 6 for the locations of these components).

Drain any solar panels, and leave the plumbing fittings or gate valves open to the atmosphere. Even in winter the heat in a solar panel can be intense, and expanding air can be as hazardous to the panels as freezing water. Store solar panels indoors if possible.

Be sure the lines to the equipment are empty or are filled with antifreeze. Remove any automatic cleaner and store it.

8. **Extra Precautions** Put a little dirt or gravel in an empty plastic bottle and leave it in the skimmer. The weight will help it stay upright in the skimmer if rain or snow refill it with water. Should the water subsequently freeze, the ice will compress the bottle, not crack the skimmer. Similarly, if you live where freezing of the pool water is likely, add an airbag to the surface of the pool underneath the winter cover. This is a vinyl balloon (use several for larger pools) that will prevent frozen water from damaging the pool walls, transferring that pressure to the airbag instead. The airbag should be held firmly on the water surface by the weight of the pool cover. When installing the cover, allow the vinyl to drape over the airbags and lie right on the surface of the water. Airbags are fitted with grommets and lines that allow you to securely position them in the pool.

During the winterizing process, look carefully for leaks in the liner and any fittings with gaskets. Small tears or gaps will become large ones when the water freezes, causing major headaches next spring. Repair any leaks before the winter freeze.

If there are exposed pipes that you suspect might still contain water, such as in the equipment area or under decks, wrap them with insulation tape (a good idea anyway to reduce heat loss when operating the system). In extremely cold areas, wrap them with electric insulation tape, available at most hardware stores on a seasonal basis. These tapes actually warm the pipe with a low-level electric current.

FAQS: REPAIRING AND WINTERIZING ABOVE-GROUND POOLS

Can I fix leaks under water?

- Yes, although the most secure patches are done on clean, dry surfaces that have been properly prepared. Small, flat areas of vinyl are the best candidates for underwater repair, while corners and areas where plumbing creates gaps are less likely to be successfully repaired that way.

Can I let my pool freeze in winter for use as a skating rink?

- No. Water expands when frozen, which can damage even steel bracing and walls. Ice also creates jagged edges that will tear vinyl liners and force open any small crevices or tears.

Can I just empty my pool and store it in winter?

- Onground pools are designed to be dismantled and stored away for the winter, but rigid-wall above-ground pools are not. In fact, you can damage components (especially the liner) by dismantling and reassembling the pool, although you can certainly do it if you are prepared to spend the time and work carefully.

If metal ladders, rails, and light fixtures cannot be removed, protect them against corrosion with a coat of petroleum jelly.

9. **Cover the Pool** Some cover is better than none because it will inhibit algae growth and keep heavy debris out of the pool (refer to Chap. 7). Sheet vinyl covers sold for pools are very inexpensive and can be held in place with clips or sand- or waterbags around the edge of the pool.

10. **Close Off Access to the Pool** Blocking access to a closed installation is an important safety precaution. Yellow caution tape strung around the pool, locked gates and fences, and extra signage will help keep the pool from becoming an inadvertent hazard and will limit your liability. At the same time, remove any deck furniture or other loose items that might be stolen, thrown, or blown into the pool. Store the automatic cleaner and water supply hose indoors. The hose should be laid out as straight as possible or coiled in large loops to avoid permanent bends.

11. **Pack Up** Take this opportunity to properly dispose of any extra chemicals or test kit reagents that won't be used during the winter

and will not be potent in the spring. Soda ash and acids are about the only water chemicals that will still be good after prolonged storage, but make sure they are packed in watertight containers and are stored in well-ventilated areas away from water or heat sources.

During the winter you still need to check on the pool. Snow or rain might raise the water level or sink the cover. Animals or heavy debris might fall in the pool and are better removed now than in the spring. Reopening the pool is essentially the reverse of the shutdown procedure, with emphasis on balancing the water before restarting the circulation system.

Not all terms in this glossary are used in the text, but because you will encounter them on product labels or in regional variances, they are included for your reference. Common words or measurement definitions not specific to the water industry are not included, but they can be found in any standard dictionary.

above-ground pool A swimming pool structure constructed of sturdy frames and panels, lined with a vinyl liner to hold the water. Also called raised pool.

acid A liquid or dry chemical that lowers pH when added to water, such as muriatic acid.

acid demand The amount of acid required (demanded) by a body of water to lower the pH to neutral (7).

acidity The quality, state, or degree of being acid.

adapter bracket The part of a pump that supports the motor and connects the motor to the pump.

air relief valve A valve on a filter that permits air to be discharged from the freeboard.

algae Airborne, microscopic plant life of many forms that grow in water and on underwater surfaces.

algicides A group of chemical substances that kill algae or inhibit their growth in water.

algistat A chemical that inhibits the growth of algae.

alkalinity The characteristic of water that registers a pH above neutral (7).

ammonia Natural substance composed of nitrogen and hydrogen that readily combines with free chlorine in water forming chloramines (weak sanitizers).

amperage (amps)   The term used to describe the strength of an electric current. It represents the volume of current passing through a conductor in a given time. Amps = watts ÷ volts.

antivortex   The property of a plumbing fitting that prevents a whirlpool effect when water is sucked through it. Used on main drain covers.

automatic gas valve   The valve that controls the release of natural or propane gas to a heater. Also called the combination gas valve.

available chlorine   *See* chlorine, free available.

backwash   The process of running water through a filter opposite the normal direction of flow to flush out contaminants.

bacteria   Any of a class of microscopic plants living in soil, water, organic matter, or in living beings and affecting humans as chemical reactions or viruses.

balance   The term used in water chemistry to indicate that when measuring all components together (pH, total alkalinity, hardness, and temperature) the water is neither scaling or etching.

balloon fitting   A plumbing connector made of rubber or flexible plastic that readily adapts to pipe of varying sizes.

barb fitting   A plumbing fitting with exterior ribs, connected by insertion into a pipe (usually flexible pipe).

base   An alkaline substance.

base plate   Square metal footing that anchors the base rail, uprights, and/or buttresses of an above-ground pool.

base rail   Metal track that forms the perimeter of an above-ground pool. The base rail is anchored to the ground by base plates, and the metal pool walls are inserted into the base rail.

bather   Any person using a pool or spa.

bather load   The number of bathers in a pool or spa over a 24-hour period.

bather occupancy   The number of bathers in a pool or spa at any one time.

bleach   Colloquial term for liquid chlorine.

bleed   To remove the air from a pipe or device, allowing water to fill the space.

blow bag   Also called a drain flush or balloon bag. A device attached to a garden hose that expands under water or uses air pressure to seal an opening, forcing the water or air into that opening.

bonding system   The wiring between electrical appliances and the ground to prevent electric shock in case of a faulty circuit. All appliances are grounded to the same wire.

bottom liner   Vinyl fabric that is laid on the ground to protect the vinyl liner of an above-ground pool (or the fabric underside of a soft-sided pool).

break point chlorination   The application of enough chlorine to water to combine with all ammonia (creating chloramines), then to destroy all chloramines, leaving a residual of free chlorine. The break point is, therefore, the point at which chlorine added to the water is no longer demanded to sanitize and is available to become residual chlorine. Some references define break point as 10 parts of chlorine for every 1 part of ammonia present in the water when the pH is between 6.8 and 7.6.

bridging   The condition existing when DE and dirt closes the intended gaps between the filter grids in a DE filter, reducing the flow rate through the filter and reducing the square footage of filter area.

bromine (Br$_2$)   A water sanitizing agent. A member of the halogen family of compounds.

Btu   British thermal unit. The measurement of heat generated by a fuel. The amount of heat required to raise 1 pound of water 1°F (when at or near 39.2°F).

buffer   A substance that tends to resist change in the pH of a solution.

buttress   The upright member of a an above-ground pool frame (especially in reference to the heavy-duty bracing used to support the long sides of oval and rectangular pools), described by architectural styles ("narrow" or "compact" or "I beam").

buttress blocks   Heavy-duty plastic pads that are placed beneath uprights to prevent them from sinking into soft ground.

calcium   A mineral element typically found in water.

calcium carbonate (CaCO$_3$)   The mineral precipitated out of water, deposited on pool and spa surfaces, the major component of scale.

calcium hypochlorite (Ca(OCl)$_2$)   A granular form of chlorine (widely produced under the brand name HTH), generally produced in a compound of 70 percent chlorine and 30 percent inert materials.

cam lock   The device that holds or releases two halves of a telepole.

cap   A plumbing fitting attached to the end of a pipe to close it completely.

cartridge   The element in a filter covered with pleats of fabric to strain impurities from water that passes through it. Generally strains out particles larger than 20 microns.

cavitation Failure of a pump to move water when a vacuum is created because the discharge capacity of a pump exceeds the suction ability.

centrifugal force The outward force created by an object in circular motion. The force that is used by water pumps to move water.

C frame A type of motor housing resembling a C. Adapts to a particular style of pump.

channeling Creation of a tube or channel in a filter media through which water will flow unfiltered. Channeling is caused by calcification of the media (hardening of the sand in a sand filter, for example) and is resolved by breaking up the sand and treating it with alum.

check valve A valve that permits flow of water or air in only one direction through a pipe.

chelating agent Chemical compounds that prevent minerals in solution in a body of water from precipitating out of solution and depositing on the surfaces of the container.

chloramine A compound of chlorine when combined with inorganic ammonia or nitrogen. Chloramines are stable and slow to release their chlorine for oxidizing (sanitizing) purposes.

chlorinator A device that delivers chlorine to a body of water.

chlorine (Cl_2) A substance made from salt that is used to sanitize water by killing bacteria. A member of the halogen family, chlorine is produced in gas, liquid, and granular form.

chlorine demand The amount of chlorine required (demanded) by a body of water to raise the chlorine residual to a predetermined level.

chlorine, free available That portion of chlorine in a body of water that is immediately capable (available) of oxidizing contaminants.

chlorine lock The term applied to chloramine formation, when ammonia is present in the water in sufficient quantity to combine with all available chlorine.

chlorine residual The amount of chlorine remaining in a body of water after all organic material (including bacteria) has been oxidized, expressed in parts per million. The total chlorine residual is the sum of all free available chlorine plus any combined chlorine (chloramine).

colorimetric The name given to a chemical test procedure where reagents are added to water and change color to reflect the presence and strength of a substance. The color is compared to a color chart to evaluate the volume of that substance. Colorimetric tests are used to detect the presence of chlorine in water.

compact buttress *See* buttress.

comparator The color chart or other device used to compare the color of a treated water sample with known values. Used in chemistry test kits.

compression fitting A plumbing connection that joins two lengths of pipe by sliding over each pipe and applying pressure to a gasket that seals the connection.

conditioner A chemical that slows the decomposition of chlorine from ultraviolet light. Conditioner, usually cyanuric acid, also helps prevent spiking of pH between high and low extremes.

control circuit A series of safety and switching devices in a heater, all of which must be closed before electric current flows to the pilot light and automatic gas valve to ignite the heater.

coupling A plumbing fitting used to connect two lengths of pipe.

cove The curved radius between the wall of the pool and the floor. Also, the packing material (either hand-formed sand or prefabricated Styrofoam) that creates a gradual slope for the liner at the point where the wall of an above-ground pool meets the ground.

CPVC Chlorinated polyvinyl chloride. The designation of plastic pipe that can be used with extremely hot water.

cyanuric acid *See* conditioner and stabilizer.

diatomaceous earth (DE) A white, powdery substance composed of tiny prehistoric skeletal remains of algae (diatoms), used as a water filtration media in DE filters. DE filters can remove particles larger than 5 to 8 microns.

dichlor *See* sodium dichloro-s-triazinetrione.

diffuser A housing inside a pump covering an impeller that reduces speed of the water but increases pressure in the system. It helps to eliminate airlock.

dog *See* tripper.

elbow The term for a plumbing connector fitting with a 90-degree bend.

electrometric A chemical test involving an electronic analysis meter.

element A filter grid. Also the electric heat-generating rod of an electric heater.

end bell   The metal housing or cap at the end of an electric motor.

erosion system   A type of chemical feeder in which granular or tablet sanitizer is slowly dissolved by constant flow of water through the device.

extrusion   A process for making plastic or metal pool components by squeezing molten material through a precut die.

female   Plumbing fittings or pipe with internal threads or connectors.

fiberoptics   Underwater lighting devices that illuminate by sending light along thin plastic cable from a remote source.

fill water   Water added to a pool or spa to replace water lost to evaporation or other reasons. Also called make-up water.

filter   A device for straining impurities from the water that flows through it.

filter cycle   *See* filter run.

filter filling   The technique used to prime a pump by filling the filter with water and allowing it to flow backward into the pump.

filter run   The time between cleanings, expressed as the total running time of the system, also called the filter cycle. Care must be taken when using the terms *run* or *cycle*, because some technicians mean the number of hours the system operates each day rather than the total time between cleanings. Regionally, one term might be used for one definition and the other for the second. I use *run* for the daily operation, *cycle* for the time between cleanings.

FIP   A female threaded plumbing fitting.

fireman's switch   An on/off control device, mounted in a time clock, that turns off a heater 20 minutes before the time clock turns off the circulating pump and motor. This allows the heat inside the heater to dissipate before shutdown.

flange gasket   The rubber sealing ring that prevents leaks between a heater and the circulation pipes.

flapper gate   The part in a check valve that swings open when water is flowing in the intended direction but swings shut when water attempts to flow backward.

flare fitting   A threaded plumbing fitting that requires a widening of the pipe at one end.

flash test   The method of dropping chemistry test reagents directly into pool or spa water rather than into a vial containing a test amount of that water.

flex connector   A coated metal pipe with threaded fittings on each end designed to bend freely for connection in tight quarters or at odd angles. Usually used to connect a gas pipe to an appliance (such as a pool heater where approved by local code) indoors, allowing the appliance to be moved without breaking the connection.

floater   A chemical feeder system whereby a sanitizer tablet is placed in the device and is allowed to float around the body of water. The tablet dissolves and the sanitizer is released into the water.

flocculate   The process of adding a chemical to a body of water which combines with the suspended particulate matter in the water, creating larger particles that are more easily seen and removed from the water. Also called a clarifying agent or coagulant.

flow rate   The volume of a liquid passing a given point in a given time, expressed in gallons per minute (gpm).

frameless pools   *See* onground pools.

freeboard   The vacant vertical area between the top of the filter media and the underside of the top of the filter.

free chlorine   Also called available chlorine. It is chlorine in its elemental form, not combined with other elements, available for sanitizing the water.

fusible link (fuse link)   A safety device located near the burner tray of a heater that is part of the control circuit. If the fuse link detects heat in excess of a preset limit, it melts and breaks the circuit to turn off the heater.

gasket   Any material (usually paper or rubber, but sometimes caulk or other pastes) inserted between two connected objects to prevent leakage of water.

gate valve   A valve that restricts water flow by raising and lowering a disc across the diameter of the pipe by means of a worm drive. *See also* slide valve.

gpm   Gallons per minute.

grid   Frame covered with fabric used as a filter media; also called septa or element.

ground fault interrupter (GFI)   A type of circuit breaker. A sensing device that determines when electricity in a circuit is flowing through an unintended path, usually to earth, creating a hazard of electrocution. The GFI detects current variations as low as 0.005 amp and breaks the circuit within one-fortieth of a second.

hardness Also called calcium hardness. The amount of dissolved minerals (mostly calcium and magnesium) in a body of water.

heater A device that raises the temperature of water using natural gas, electricity, propane, solar, or mechanical energy for fuel. To be called a heater, the device must convert at least 70 percent of its fuel into heat (no more than 30 percent lost in venting). Over 80 percent efficiency, the device is typically called a boiler.

heat exchanger The copper tubing in a heater through which water flows. The water absorbs rising heat that is generated from the burner tray below.

heat riser A metal pipe plumbed directly to the heater to facilitate dissipation of heat before plumbing with PVC. Also called a heat sink.

high-limit switch A safety device used in the control circuit of heaters. When the high-limit switch detects temperatures in excess of its preset maximum, it breaks the control circuit to shut down the heater.

high-rate sand filter A filter using sand for the filtration media designed for flows in excess of 5 gpm but less than 20 gpm (less than 15 gpm in some codes) per square foot. Strains impurities larger than 50 to 80 microns.

horsepower (hp) The standard unit of measurement that denotes the relative strength of a mechanical device. One horsepower equals 746 watts or the power required to move 550 pounds 1 foot in 1 second.

hose bibb (also bib) The faucet to which a garden hose is attached.

hydrochloric acid (HCl) Muriatic acid.

hypochlorous acid (HOCl) A form of free available chlorine resulting from the solution of an active chlorine compound added to water.

I-beam buttress *See* buttress.

impeller Rotating part of a pump that creates centrifugal force to create suction. The impeller is said to be closed if it is shrouded (covered) on both sides of the vanes, or semiopen if it is shrouded on one side, while the interior surface of the volute creates a partial shroud on the other side.

inlet   *See* return.

intermittent ignition device (IID)   The electronic control and switching device used in electronic ignition heaters to operate the control circuit and automatic gas valve. Often called the brainbox.

joint stick   A paste used on threaded plumbing to prevent leaks that is available in the form of a crayon-like stick for easy application.

keyed shaft   The shaft of a motor that has a groove for securing setscrews of the shaft extender. Used with specific designs of pumps.

kilowatt   One thousand watts of electrical power. Electricity is sold by the kilowatt-hour, meaning a certain fee is charged for every 1000 watts delivered per hour.

laterals   The horizontal filter grids at the bottom of a sand filter, installed in the underdrain.

lazy flame   A natural or propane gas flame in the burner tray of a pool or spa heater that burns in slow, wavering licks or stuttering, rather than the normal strong, clear blue flame burning straight upward.

leafmaster   A brand name; leafmaster is a term applied to any device that vacuums large debris from a pool by means of water pressure created with a garden hose.

leaf rake   A large open net secured to a frame that attaches to a tele-pole that is used to skim debris from the surface of the water.

legs   *See* upright.

main drain   The suction fitting located in the lowest portion of a body of water. The principal intake for the circulation system.

main valve   The flow control device in the combination gas valve of a pool or spa heater that regulates the flow of gas to the burner tray.

male   Any plumbing fitting or pipe with external threads or connectors.

manifold   An assembly or component that combines several other components together. A pipe fitting with several lateral outlets for connecting one pipe with others.

media   Any material used to strain impurities from water that passes through it. DE and the fabric covering a cartridge are both examples of filter media.

micron   A unit of measurement equal to 0.000001 meter or 0.0000394 inch. For example, a grain of table salt is approximately 100 microns in diameter.

MIP   A male threaded plumbing fitting.

mission clamp   A trade name. A mission clamp is a rubber connector hub-type fitting, secured with a stainless steel clamping band, used in plumbing to join two lengths of pipe of similar or dissimilar diameter.

muriatic acid   Also called hydrochloric acid. This chemical is the most commonly used substance for reducing pH and total alkalinity in water.

narrow buttress   *See* buttress.

neutral   The pH reading at which the substance being measured is neither acidic nor alkaline. Neutral pH is 7.0.

nipple   A short length (less than 12 inches) of pipe threaded at each end. If the nipple is so short that the entire length is threaded, it is called a close nipple.

no-hub connector   A rubber (or neoprene) plumbing fitting used to connect two lengths of pipe, attached with hose clamps or other pressure devices.

onground pools   One-piece swimming pools of various fabrics (significantly thicker than the vinyl liner of an above-ground pool) that are self-standing units, including those that are inflated. Also called frameless pools.

open loop   A type of solar heating system that circulates the water being heated from the pool or spa, through the solar panels, and back to the pool or spa.

O-ring   A thin rubber gasket used to create a waterproof seal in certain plumbing joints or between two parts of a device, such as between the lid and the strainer pot on a pump.

ORP   Oxidation reduction potential. A unit of measure of sanitizer ability in water, measured with an electronic ORP meter.

OTO   Orthotolidine. The test reagent used in detecting the presence of chlorine and, by the resulting color of the OTO, the amount of chlorine, expressed in parts per million.

ozonator   A device using electricity and oxygen to create ozone and deliver it to a body of water for sanitizing purposes. Ozone is produced in two ways for pool use. One method, called the corona discharge method, is to shoot an electrical charge through oxygen and water, creating ozone. The other method is to electrify freon (the gas used in air conditioners and refrigerators) to produce ultraviolet light. In both cases, the production happens in a chamber in an ozonator built into the circulation system.

ozone (O_3)   Three atoms of oxygen, creating a colorless, odorless gas used for water sanitation.

pH   The relative acidity or alkalinity of soil or water, expressed on a scale of 0 to 14, where 7 is neutral, 0 is extremely acidic, and 14 extremely alkaline.

phenol red   (phenolsulfonephthalein) The most widely used chemical reagent used to measure the pH in a sample of water.

piezo   An ignition unit that creates a spark to ignite the pilot burner in a gas heater.

pilot   The small gas flame that ignites the burner tray of a heater.

pilot generator   The device that converts heat from the pilot light into electricity to power a control circuit on a heater. Also called a power pile or thermocouple.

pipe dope   A paste used to prevent leaks in threaded plumbing.

pipe run   A length of pipe between two valves, connectors, or pieces of pool equipment.

plug   A plumbing fitting used to close a pipe completely by inserting it (slip or threaded) into a female fitting or pipe end.

port   An opening, as in a discharge port being the opening through which water flows out of a pipe or system.

portable pool   A pool that can be readily disassembled, moved, and reassembled. Both above-ground and onground pools are portable.

positive seal   The type of multiport valve that directs all of the water flow to one direction, allowing no flow in the other lines. A non-positive seal directs most of the water to one line, allowing some water to bypass into the other lines.

potassium monopersulfate   An oxidizing chemical used to sanitize water. Used to supersanitize without using chlorine. Also catalyzes bromine.

ppm   Parts per million. The measurement of a substance within another substance; for example, 2 ounces of chlorine in 1 million ounces of water would equal 2 ppm.

precipitate   An insoluble compound formed by chemical action between two or more normally soluble compounds. When water can no longer dissolve and hold in solution a compound, it is said to precipitate out of solution.

precoat   The process of applying DE to grids in a DE filter after cleaning but before restarting normal circulation and filtration.

pressure gauge   A device that registers the pressure in a water or air system, expressed in pounds per square inch (psi).

pressure sand filter   A type of sand and gravel filter in which the water is strained through the filter media under pressure (as opposed to a free-flow filter). *See also* high-rate sand filter.

pressure switch   A safety device in a heater control circuit that senses when there is inadequate water pressure (usually less than 2 psi) flowing through a heater (which might damage the heater) and breaks the control circuit, thereby shutting down the heater.

prime   The process of initiating water flow in a pump to begin circulation by displacing air in the suction side of the circulation system.

psi   Pounds per square inch.

pump   A mechanical device driven by an electric motor that moves water.

PVC   Polyvinyl chloride. The type of plastic pipe and fittings most commonly used in pool and spa plumbing.

raised pool   *See* above-ground pool.

reagent   A liquid or dry chemical that has been formulated for water testing. A substance (agent) that reacts to another known substance, producing a predictable color.

reducer   An external plumbing fitting that connects two pipes of different diameter.

residual   The amount of a substance remaining in a body of water after the demand for that substance has been satisfied.

retaining rod   The metal rod in the center of certain filters on which is attached a retainer ring to hold grids in place.

retainer   The plastic disc that fits over the top of a set of filter grids to hold them in place with the aid of a retainer rod.

return   The line and/or fitting through which water is discharged into a body of water. Also called an inlet.

riser   *See* heat riser. Also refers to any vertical run of pipe.

roll-forming   A process for making metal above-ground pool walls from coiled sheets of aluminum or steel.

rotor   The rotating part of an electric motor that turns the drive shaft that itself is driven by magnetism produced between it and the stator. Also the diverter in the backwash valve of certain types of filters.

run   Any horizontal length of pipe. *See also* filter run.

sand filter   A filtration device using sand as the filter (straining) media. *See also* high-rate sand filter.

sanitizer   Any chemical compound that oxidizes organic material and bacteria to provide a clean water environment.

saturation   The point at which a body of water can no longer dissolve a mineral and hold it in solution.

scale   Calcium carbonate deposits that form on surfaces in contact with extremely hard water. Water in this condition is said to be scaling or precipitating.

seal   A device in a pump that prevents water from leaking around the motor shaft.

seal plate   The component in a pump in which the seal is situated.

separation tank   A container used in conjunction with a DE filter to trap DE and dirt when backwashing.

skidpack   A metal or plastic frame onto which is mounted the equipment needed to operate a portable pool, usually a pump and motor, filter, heater, and control devices.

skimmer   A part of the circulation system that removes debris from the top of the water by drawing surface water through it.

slide valve   A guillotine-like plumbing device that restricts or shuts off the flow of water in a line. In essence, a gate valve that slides a disc up or down across the flow in the pipe. *See also* gate valve.

slip fitting   A plumbing fitting that joins to a pipe without threads, but that slides into a prefitted space.

slurry   A thin, watery mixture.

soda ash (Na_2CO_3)   Sodium carbonate. A white powdery substance used to raise the pH of water.

sodium bicarbonate ($NaHCO_3$)   A chemical used to raise the pH and total alkalinity of water. Also called baking soda, bicarb, and bicarbonate of soda.

sodium bisulfate ($NaHSO_4$)   A chemical compound used to lower the pH and total alkalinity of water (dry acid).

sodium dichloro-s-triazinetrione ($C_3N_3O_3Cl_2Na$)   Dichlor. A granular, stabilized form of chlorine sanitizer, generally about 60 percent available.

sodium hypochlorite ($NaOCl$)   A liquid solution containing approximately 15 percent chlorine.

sodium thiosulfate   A chemical used to neutralize chlorine in a test sample prior to testing for pH, without which a false reading might result.

soft-sided/framed pool A variety of above-ground pool that uses a heavy-duty vinyl fabric bag to hold the water, braced by a series of metal supports, but does not use rigid wall panels.

soft water Water that is very low in calcium and magnesium (less than 100 ppm).

solar panel A metal, glass, or plastic enclosure, usually 4-by-8 feet by a few inches thick, through which water flows absorbing heat from the sun. The basic component of a solar heating system.

square flange A casing style of certain motors, designed to adapt to certain pumps.

stabilizer Any compound that tends to increase water's resistance to chemical change. *See also* conditioner.

stack Refers to the vent pipe of a heater. A heater installed indoors requires such a stack, while one installed outdoors uses a flat-top vent and is called stackless.

standing pilot A heater ignition device in which the gas flame (to ignite the main burner) is always burning.

stator The stationary part in an electric motor that contains the wound wires (winding) that generates magnetism to drive the rotor.

strainer basket A plastic mesh container that strains debris from water flowing through it inside the strainer pot.

strainer pot The housing on the intake side of a pump that contains a strainer basket and serves as a water reservoir to assist in priming.

street fitting Any plumbing fitting that has one male end and one female end.

submersible pump A pump and motor that can be submerged to pump out or recirculate a body of water. Also called a sump pump.

superchlorinate Periodic application of extremely high levels of chlorine (in excess of 10 ppm) to completely oxidize any organic material in a body of water (including bacteria) and leave a substantial chlorine residual. Procedure performed to sanitize elements in water that might resist normal chlorination. Also called shocking or shock treatment.

Teflon tape A thin fabric provided on a roll used to coat threaded plumbing fittings to prevent leaks.

telepole A metal or fiberglass pole that extends to twice its original length, the two sections locking together. The telepole is used with most pool and spa cleaning tools.

T fitting   A plumbing fitting shaped like the letter T that connects pipes from three different sources.

therm   The unit of measurement you read on a gas bill; is 100,000 Btu/hour of heat.

thermal overload protector   A temperature-sensitive switch on a motor that cuts the electric current in the motor when a preset temperature is exceeded.

thermostat   A part of the heater control circuit. An adjustable device that senses temperature and can be set to break the circuit when a certain temperature is reached. It then closes the circuit when the temperature falls below that level.

three-port valve   A plumbing fitting used to divert flow from one direction into two other directions.

time clock   An electromechanical device that automatically turns an appliance on or off at preset intervals.

titration   A chemical test method to determine the amount of a substance in a sample of water.

total alkalinity   The measurement of all alkaline substances (carbonates, bicarbonates, and hydroxides) in a body of water.

total dissolved solids (TDS)   The sum of all solid substances dissolved in a body of water, including minerals, chemicals, and organics.

trichloro-s-triazenetrione ($C_3N_3O_3Cl_3$)   Trichlor. A dry form of stabilized chlorine, produced in granular or tablet form at around 90 percent available chlorine.

tripper   The small metal clamp that fits in the clock face of a time clock to activate the clock (on/off) at preset times. Also called a dog.

trisodium phosphate   A detergent used to clean filter grids and cartridges, it breaks down oils that acid washing alone cannot. Commercially sold as TSP.

turbidity   Cloudiness.

turnover rate   The amount of time required for a circulation system to filter 100 percent of the water in a particular body of water.

U brace   *See* upright.

union   A plumbing fitting connecting two pipes by means of threaded male and female counterparts on the end of each pipe.

upright   The vertical column used to stiffen metal walls in above-ground pools. Also called leg, U brace, or vertical tube.

upright straps   Fabric panels attached to soft-sided/framed pools that connect the uprights to the base of the pool.

vacuum   A device used to clean the underwater surfaces of a pool or spa by creating suction in a hose line.

valve   A device in plumbing that controls the flow of water.

vertical tube   *See* upright.

vinyl liner   The plastic fabric bag that is draped inside above-ground pool walls to make the vessel hold water.

volute   The volute is the housing that surrounds the impeller and diffuser in a pump, channeling the water to a discharge pipe.

weir   The barrier in a skimmer over which water flows. A floating weir raises and lowers its level to match the water level in a pool. Another type is shaped like a barrel and floats up and down inside the skimmer basket.

winterizing   The process of preparing a pool to prevent damage from freezing temperatures and other harsh weather conditions.

REFERENCE SOURCES AND WEBSITES

Virtually every equipment and supply manufacturer produces valuable literature with repair and maintenance information, technical guidelines, and exploded diagrams of their products. Collect literature from websites, from your supply house, and anywhere else you can.

The following publications were used for reference in preparing this book:

Tamminen, Terry. *The Ultimate Pool Maintenance Manual*. New York: McGraw-Hill, 2001.

Water Analysis Handbook. Loveland, Colo.: Hach Company, 1992.

Manas, Vincent T. *National Plumbing Code Handbook*. New York: McGraw-Hill, 1957.

Uniform Swimming Pool, Spa and Hot Tub Code. International Association of Plumbing and Mechanical Officials, 1988.

Taylor, Charlie. *Everything You Always Wanted to Know about Pool Care but Didn't Know Where to Ask*. Chino, Calif.: Service Industry Publications, 1989.

Wood, Robert W. *Home Electrical Wiring Made Easy*. 2d ed. Blue Ridge Summit, Pa.: TAB Books, 1993.

Websites

The Web makes all companies seem equal, but, in fact, many are international conglomerates while others are very local (one above-ground pool maker sells in the Quad Cities only while another sells in Amish country only!). That's good for finding a unique pool made by local crafters, but it complicates shopping if you choose a pool that costs twice the price because of shipping (and nowhere to see a sample). Pool equipment can easily be shipped, but when surfing the Web for an onground or above-ground pool, you might want to start with the dealer locator section to be sure you can get local sales and service.

Many of the websites in this section feature both pools and pool equipment. Almost all sites offer FAQs about above-ground pools, water maintenance, and links to other sites. My notes highlight especially good aspects of these sites, but every one of them has more to offer than could possibly be indexed here.

It's also worth mentioning that websites change frequently, while others should be updated more often than they are. For example, one major manufacturer of excellent above-ground pools warns that you can buy their products only through authorized dealers and directs you to a dealer locator, but up to the time of this book's publication, when you click on that site, you are linked to a wedding planner of the same name!

The moral of the story is that every company has a website. The websites listed below were used in the writing of this book, but this list is not necessarily complete in terms of the Web-based references that might help you. Look around and even revisit sites frequently to see how they might have changed. Also look at links on each website because many sites are themselves great references to other helpful sites. Have fun!

abovegroundpools.com	Cast resin pools
abovegroundpoolslide.com	Just what the name implies
aqua-flo.com	User-friendly pool and spa equipment specs, photos, pdf format downloads
baquacil.com	Dozens of products for pool chemistry; good FAQs with details on biguanicides
cantar.com	Good starting point for above-ground pools
corneliuspools.com	Cool animation!
crestwoodpools.com	Useful information on pressure-treated lumber and other wooden pool aspects; good photo gallery and technical manuals
delairgroup.com	Great above-ground pool photos and "buyers guide"
doughboy-pools.com	The original above-ground pool

eren.doe.gov/rspec	Very creative, useful site about saving water and energy in pools and spas
eTrevi.com	Canadian manufacturer of steel-walled round pools
fafco.com	Step-by-step layout of solar pool heating, lots of tech data
foxpools.com	Sells the "Amish Onground Pool"; one model—metal wall rectangle with deck
gpspool.com/dealers	Good list of manufacturers with technical downloads
intermatic.com	Timeclocks and more
jacuzzibros.com	Great online parts catalog; sells a lot more than jetted tubs—pumps, filters, auto cleaners, more
jandy.com	Good selection of pool and spa equipment, especially Laars heaters; good tech info and training seminars
kayakpools.com	Aluminum pools; rectangle only (they argue that it's better for swimming and water sports); all with decks; website simple to navigate, showing products, landscaping, decks, etc.; "katalog" has all manner of equipment and supplies for sale online
kd.com	Great "university" of water maintenance topics; good description of onground pools, including assembly
medallionPools.com	Lots of FAQs
michpool.com	Pool and deck packages
mr-pool.com	Good place to start for basic FAQs; most popular products; nice list of links
muskin.com	Portable pools; great movie intro and user-friendly site

northwestwholesale.com	Good links and technical information; nice photos, especially of installation of above-ground pool
ovationpools.com	Portal to several manufacturers
pentairpool.com	Very useful "answer pool" section, including a neat pool capacity calculator; good literature and downloads
polarispoolsystems.com	Good site for automatic pool cleaners, with FAST diagrams and photo downloads
poolproducts.com	Great tips on replacement liners (very detailed, by shape); good directory of tips, products, pools/kits
poolspanews.com	Great searchable site with lots of tech help and links
poolsupplies.com	Great site representing variety of above-ground pool types, supplies, info; sells FantaSea pools
pooltoolco.com	Great specialty tools
raypak.com	Easy-to-download spec sheets on pool heaters; training seminars
seaspraypools.com	Good installation downloads
secardpools.com	Basic pools and supplies
sentrypool.com	Pools and water features
sharkline.com	Excellent how-to downloads; terrific selection of metal and soft-sided pools; good photos, including actual pool owner photos
sofpool.com	Very user friendly site; one photo says it all—the entire pool in a duffel bag with the caption "Winterizing"!
solarpools.com	Same site as poolsupplies.com

solar-tec.com	Good site for intro to solar; pricing, all user-friendly
splashpools.com	Excellent technical downloads
splashsuperpools.com	Great product selection, photos, support manuals, options
sta-rite.com	Good site with easy-to-get owner's manuals and a great chart that tells you how long each will take to download at your modem speed
swimartesian.com	Stainless-steel pool maker
swimmingpools.com	Lots of good pool "how-to's" including vinyl liner installation
swimnplay.com	All the basics
technobois.com	Unique wooden above-ground pools; great explanations of the architecture, materials, and photos of the pools
voguepools.com	Good photos of decks, landscape ideas
wilkespools.com	One model, but all components included (deck, ladders, etc) and well explained
zodiacpools.com	Quick; good product info on inflatables

INDEX

ABOUT THE AUTHOR

Terry Tamminen is a leading expert on pools and spas. The owner/operator of a Malibu, California, pool/spa service business with a star-studded clientele, he has provided technical advice on large-scale water chemistry and pool technology projects for over 25 years. With Robert F. Kennedy, Jr., Mr. Tamminen founded numerous Waterkeeper programs to protect our nation's rivers and coastlines. He currently serves as Secretary of the California Environmental Protection Agency and is also author of *The Ultimate Pool Maintenance Manual,* published by McGraw-Hill.